THE
SMITHSONIAN
NATIONAL
GEM
COLLECTION

UNEARTHED

THE SMITHSONIAN NATIONAL GEM COLLECTION

UNEARTHED

SURPRISING STORIES BEHIND THE JEWELS

JEFFREY EDWARD POST

Abrams, New York

Contents

The Story of a Gem 6

A Storied Collection 11

A Dozen Notable Gems 109

Gem Families 123

The Hope Diamond 165

The Story of a Gem

When a mineral crystal is cut and polished into a gem, it is transformed not just in appearance but in perception; a gem assumes a value—and a story. Where did it come from, who owned it, is there a mysterious past, does it carry mystical powers, or perhaps a curse? Gems represent stability; unlike living things, their beauty is undiminished with time. The stone that once adorned a queen or a movie star might now be set in a dinner ring and, if not destroyed, will sparkle as brightly thousands of years from now. To paraphrase a successful marketing slogan: A gem is forever. Gems accumulate history, and in many cases that provenance, the story, contributes as much to the perceived value of a gem as its rarity, size, and beauty. The gem might be a passive observer in its history, or help make it. Every good jeweler knows that it is the romantic story that helps to sell the stone, even if sometimes one must be made up.

But what makes gems valuable? A mineralogist might emphasize the rarity of gem-quality crystals in the earth or wax on about the perfection of their atomic structures; a gemologist can expertly describe their clarity, color, and the quality of the cut. And, no doubt, they appear more special when mounted into exquisite settings that enhance their brilliance and beauty. Like a magnificent work of art, a gem can be valued simply because it gives us pleasure. But the fact is, gems have little intrinsic value: We cannot eat them, they do not cure illnesses, and most do not make particularly good weapons. We can get along just fine without them. The real worth of gems are the values that we and our ancestors have agreed to assign them. Throughout human history, people have used stones for adornment and currency. We have long been fascinated by certain colorful and sparkly stones; they are history's preeminent symbols of wealth and power, and even now suggest a certain social status. Gems are an important form of portable wealth; some have perhaps the highest value per volume of any material known. Gems are easily hidden and transported, and because their values are universally recognized, they can be readily converted to cash anywhere in the world. And they are perhaps the only investments that can be enjoyed in jewelry and convey a certain status to the wearer.

Of all the objects in the National Museum of Natural History, gems perhaps best, and uniquely, provide an intersection of natural science, human history and culture, romance, the skill and creativity of artists and craftspeople, the allure of immense value, and the awe of stunning beauty. It is no wonder, then, that the gem and mineral galleries are among the most popular of all Smithsonian exhibitions, appealing to all genders and ages. But of course, we are mere latecomers to a gem's story; every gemstone was cut from an exceptional mineral crystal that grew in the earth millions or even billions of years ago. In many ways, the real miracle behind every gem is that the crystal survived eons of geological forces before it was mined and faceted, and that alone makes each one special.

One of the world's greatest collection of gems is in the Smithsonian National Museum of Natural History. It includes iconic stones such as the Hope Diamond, Star of Asia Sapphire, Bismarck Sapphire, Hooker Emerald, and Blue Heart Diamond. It is a collection not only of rare and beautiful jewels, but also of the stories of the people who once owned or were associated with them—royalty, movie stars, the rich and famous, and "ordinary" folks. Did you know that New York adman Rosser Reeves donated his great star ruby because he couldn't resist the alliteration Rosser Reeves Ruby, or that Polly Logan gave her huge sapphire in part because it reminded her of her unfaithful previous husband? The Napoleon Diamond Necklace was sold by swindlers, resulting in an Austrian archduke being tried for the crime in New York City, and the Countess Mona von Bismarck, who gave the spectacular Burmese sapphire necklace, was the daughter of a Kentucky horse trainer. This book is about the great gems and jewelry pieces in the Smithsonian National Gem Collection, but mostly it is about their stories—who are the people who owned them, who donated them, and why? The stories are forever bonded with the stones, an inseparable intertwining of people's lives and earth's rarest treasures. Gems are portals that offer fascinating glimpses into the lives of the famous and powerful, and of those who were neither. For some people who happened into an association with a great jewel, it is perhaps only because of that gem that their story will be remembered—immortalized in stone. We currently share a moment with these gems; we can research and document their pasts, but only speculate about their futures. Their stories are ongoing, and perhaps only beginning.

PAGE 2
Yogo Flower Brooch, page 131

PAGE 4
Dom Pedro Aquamarine, page 94

OPPOSITE
Smithsonian diamonds! The 253.7-carat Oppenheimer Diamond crystal (top) with the 30.62-carat Blue Heart Diamond (center), surrounded by a selection of faceted diamonds in the Smithsonian National Gem Collection.

The United States National Gem Collection

When Englishman James Smithson left his fortune in 1834 to the then fledgling United States of America for the establishment of the Smithsonian Institution, he also bequeathed his collection of more than ten thousand mineral specimens. Smithson was a chemist and mineralogist, and the zinc carbonate mineral smithsonite, which he first described as a distinct mineral, was named after him. Unfortunately, his complete collection and its documentation were destroyed in a fire in the Smithsonian Castle in 1865. It is not known whether Smithson's collection included cut gems, but almost certainly there were minerals that are commonly used as gems.

The present National Gem Collection at the Smithsonian Institution has grown primarily as an accumulation of gifts, mostly from individuals, to the people of the United States and the world. In 1886, the noted gemologist George F. Kunz described the collection, stating, "Although a mere beginning, it is the most complete public collection of gems in the United States." The collection received a major boost in 1894 with the bequest of 1,316 precious stones from Philadelphia publisher Dr. Isaac Lea's extensive collection, gifted by his daughter, Frances Lea Chamberlain. This was followed in 1896 by a notable gift of gems from Mrs. Chamberlain's husband, Dr. Leander T. Chamberlain. In 1897, Dr. Chamberlain became honorary curator of the collection and added several fine gems and later bequeathed a modest endowment for gem acquisition.

The next significant event in the growth of the gem collection was the donation in 1926 of the superb mineral collection of Colonel Washington A. Roebling by his son John A. Roebling. Colonel Roebling is probably best known as the builder of the Brooklyn Bridge. The Roebling gift included many fine rough and cut gems and was accompanied by an endowment that continues to support the collection.

The single most important event that established the worldwide reputation of the National Gem Collection was the gift on November 10, 1958, by Harry Winston of the renowned Hope Diamond. Not only did this iconic gem become a major attraction, but its arrival triggered a series of major gifts—such as the Blue Heart Diamond, Napoleon Diamond Necklace, and Marie Louise Diadem from Marjorie Merriweather Post; the Logan Sapphire from Polly Logan; the Hooker Emerald from Janet Annenberg Hooker; the Carmen Lúcia Ruby from Peter Buck; and the recent Whitney Flame Topaz from Coralyn Wright Whitney—that have built the National Gem Collection into one of the greatest public displays of gemstones in the world. New annual gifts assure the collection's continued growth.

What Is a Mineral, Crystal, or Gem?

Minerals are naturally occurring chemical compounds that are the basic building blocks of the solid earth. All rocks are made of minerals. For example, the dark and light grains that make up the rock granite are primarily quartz and one or more feldspar minerals. Crystals are the natural form of most minerals, meaning they are constructed of atoms that are locked into precise symmetric patterns that are repeated in an orderly way in three dimensions billions of times. At the ideal temperature with a plentiful supply of the right kinds of atoms, large, perfect crystals can grow in cavities and cracks in the earth from magma or hot water solutions. Gems are mineral crystals that have been cut and polished by skilled craftsmen into objects of beauty. Theoretically, any of the nearly 5,600 known minerals can be used as gems, but only a small fraction combine desirable qualities such as attractive color, clarity, brilliance, and durability with sufficient availability to be marketable. Approximately sixteen minerals account for most commercially available gems:

Diamond	Elbaite
Corundum	Garnet family
Beryl	Zoisite
Topaz	Feldspar family
Quartz	Chrysoberyl
Spinel	Opal
Spodumene	Jadeite
Zircon	Olivine family

RIGHT, ABOVE

The minerals in the feldspar family make up more than half of the earth's crust. Occasionally, these common minerals form crystals that exhibit a reddish to golden sheen called schiller, resulting from light reflecting off numerous tiny copper or hematite (iron oxide) flakes scattered within the stones. Copper in Oregon labradorite sunstone also produces shades of green and red, as in this 32.3-carat gem carving from Plush, Oregon, by Naomi Sarna. Donated to the National Gem Collection by Doug and Robin Malby in 2016.

RIGHT

Jadeite is commonly found in Burma (also known as Myanmar) as stream-worn cobbles with a weathering rind. The miners notch the stones to assess the quality of the jadeite. The Chinese markings record dealer and inventory information (height 24.6 centimeters, or 9.7 inches).

A Storied Collection

The Legacy of Marjorie Merriweather Post

Growing up as a member of a large Post family, I often wondered about the Post name on the cereal boxes on our breakfast table. It certainly was not anyone from our branch of that family tree. Little could I have imagined that I would become curator of a gem collection that owes much of its success to one of these "other" Posts. The gift of the Hope Diamond in 1958 by Harry Winston instantly made the Smithsonian Institution a destination for people wanting to see the famous gem, but it was a series of donations of spectacular gems and jewelry pieces in the 1960s by Marjorie Merriweather Post that helped build a modest display of gems into a world-class National Gem Collection.

Marjorie Merriweather Post was the only child of Ella Merriweather and Charles William Post. C. W. Post was a pioneer in the cereal industry and founder of the Postum Cereal Company. When he died in 1914, his daughter Marjorie became a major stockholder in what would become General Foods, and at age twenty-seven she was one of the wealthiest women in the United States. In the early 1920s, women's fashions and elaborate jewels celebrated newfound freedoms. Great gems and spectacular jewelry brightened elite social gatherings. Ms. Post made her first major jewelry purchases with her husband, financier E.F. Hutton, whom she married in 1920. From the 1920s through the 1960s, she was one of Cartier's most important American clients; the Post Emerald Necklace, Maximilian Emerald, and Marie Antoinette Diamond Earrings, which are now part of the Smithsonian collection, were purchased by Ms. Post at this time.

From 1936 to 1938, Ms. Post's then husband, Washington, DC, lawyer Joseph Davies, served as ambassador to the Soviet Union. Their time in Moscow introduced Ms. Post to the gem and jewelry treasures of the Russian tsars, resulting in the incredible collection of Fabergé and other Russian-inspired objects that remain at the Hillwood Estate, Museum & Gardens. This experience encouraged an interest in exquisite jewels with royal or aristocratic provenance that guided many future acquisitions. After World War II, many European aristocrats were financially devastated and forced to sell the family jewels to survive. Several great jewelry houses acquired the antique pieces primarily for the gems, which could be recut and set into contemporary styles. Fortunately, Ms. Post's collecting interests saved at least a couple of historic jewelry pieces; she purchased from Harry Winston the diamond necklace given by Emperor Napoleon to Empress Marie Louise in 1811. Ms. Post also purchased from Van Cleef & Arpels, and later donated to the Smithsonian, a diadem that was a wedding gift from Napoleon to Marie Louise.

In the 1960s, Ms. Post started thinking about leaving her home, Hillwood, to the Smithsonian Institution, and she began donating several jewelry pieces to the National Gem Collection. At an event in 1969, Smithsonian secretary S. Dillon Ripley recalled how in 1964, Ms. Post appeared at the Smithsonian Castle accompanied by several friends and her granddaughter. She opened her shopping bag and revealed an "irreplaceable collection of jewels and lace." Included in the cache were the Blue Heart Diamond, Maximilian Emerald, and Marie Antoinette Diamond Earrings. In addition to her many generous donations, Ms. Post also encouraged her friends to give jewelry pieces to the growing gem collection. In a letter to Anna Thompson Dodge, wife of Horace Dodge, cofounder of Dodge automobiles, she wrote: "The Smithsonian would be so grateful, and it would give such pleasure to the many thousands who visit the museum." Hillwood was transferred to the Smithsonian Institution in December 1968, and many of her donated items were given on the stipulation that they be returned to her home for display upon her death for formation of a house museum operated by the Smithsonian. After her death in 1973, the Smithsonian determined it would not have the resources necessary to convert Hillwood into a museum. Accordingly, Hillwood and certain collection items were returned to the Post Foundation in 1976.

A 1968 inventory of Ms. Post's jewelry collection listed 208 pieces. Most were kept in a safe near her dressing room, so as to be easily accessible for any outfit at any time. But as Liana Paredes, chief curator at the Hillwood Estate, Museum & Gardens, said of Ms. Post: "She saw jewelry not only as objects for personal adornment, but also as works of art worthy of display." The fortunate results of that philosophy are many of the iconic and beautiful gems and jewelry pieces that thrill the millions of annual visitors to the Smithsonian National Gem Collection gallery.

PAGE 10
Marie Louise Diadem, page 18, with original emeralds

OPPOSITE
Portrait of Marjorie Merriweather Post in presentation court dress wearing the Marie Antoinette Diamond Earrings and the Maximilian Emerald Ring, 1929. (Giulio de Blaas; Hillwood Estate, Museum & Gardens)

ABOVE
Marjorie Merriweather Post delivering the Napoleon Diamond Necklace to Smithsonian Institution secretary Leonard Carmichael (left), 1962.

Marie Antoinette Diamond Earrings

When Marjorie Merriweather Post purchased these impressive diamond earrings from Pierre Cartier in October 1928, he provided documentation that they had once belonged to Queen Marie Antoinette of France. The large diamond drops were in their original silver settings that had gold linkages and were decorated with old-mine-cut diamonds in scrollwork. Later, Cartier replaced the tops of the earrings with triangular diamonds set in platinum. In 1959, Harry Winston Inc. mounted the large diamonds into platinum and diamond replicas of the original settings. The drops were made detachable, and round diamonds and diamond-set links were added so that Ms. Post could attach them to a necklace with a 13.95-carat triangular center diamond. In November 1964, Ms. Post's daughter, Eleanor Barzin, donated the earrings, along with the original settings, to the Smithsonian Institution. The diamonds and links added to the drops by Winston's were removed prior to their donation, but they retain the Cartier tops and the modern platinum and diamond settings. The large pear-shape diamonds weigh 20.34 and 14.25 carats, respectively.

It is not difficult to believe that these impressive diamond earrings must once have been owned by someone important, but was that person in fact the ill-fated Queen Marie Antoinette, who was executed by guillotine during the French Revolution? This question has been researched and debated from the time the earrings were acquired from Cartier in 1928. The abundance of circumstantial evidence is intriguing, but is it enough to make a case? You decide!

In 1774, when Marie Antoinette reigned as queen, it was a time of royal extravagance, court intrigue, and popular unrest that culminated in the French Revolution in 1789. The royal family was arrested in October of that year, and the monarchy was abolished in 1792. Marie Antoinette was executed in October 1793.

Germain Bapst's 1889 history of the French Crown Jewels mentions that King Louis XVI gave Marie Antoinette earrings set with pear-shape diamonds hanging from four diamonds at the posts, which she is reported to have worn constantly. In a footnote (translated from French), Bapst quotes the queen's lady-in-waiting, Madame Campan: "Mr. Boehmer, court jeweler, had assembled six large diamonds on order of Louis XV for [Madame] Du Barry but were not given before the king's death. So, Mr. Boehmer set two as earrings and offered them to the new queen. Marie Antoinette could not afford the four hundred thousand livres and turned them down. The king increased her allowance and she obtained them." In her book *Spectacular*, about the gems and jewelry of Marjorie Merriweather Post, Liana Paredes points out that the French term "girandoles," used by Bapst to describe the earrings, refers to a style of earring with a large gem surrounded by additional dangling stones, and she concluded these earrings, therefore, do not match the pair owned by Ms. Post.

LEFT
Portrait of Queen Marie Antoinette wearing pear-shape drop diamond earrings. Élisabeth Louise Vigée Le Brun, 1788. Versailles.

OPPOSITE
The Marie Antoinette Diamond Earrings as donated to the Smithsonian Institution by Marjorie Merriweather Post's daughter Eleanor Barzin in 1964. The triangular diamond tops were designed by Cartier. The diamonds are in settings made by Harry Winston Inc. in 1959 that are replicas of the originals.

However, in her memoirs, Madame Campan recorded that although the queen returned all jewels belonging to the Crown to the commissioner of the assembly, she notes: "Her Majesty retained nothing but a suite of pearls and a pair of earrings, composed of a ring and two drops, each formed of a single diamond. These earrings and several fancy trinkets, which were not worth the trouble of packing up, remained in Her Majesty's chest of drawers at the Tuileries."

It seems reasonable to conclude that Queen Marie Antoinette had at least one pair of diamond earrings, and perhaps more, whose description matches those in the Smithsonian collection. Unfortunately, there is no information as to their whereabouts immediately after the queen's death (they were considered her personal property and were not listed in inventories of the crown jewels). Many of her personal jewels, perhaps including the diamond earrings, were sold by her daughter Marie Thérèse in 1799.

In 1874, a reference to Marie Antoinette earrings appears in a sales catalog for an auction in Geneva that lists two pairs, one set with pearls and the other with large drop diamonds weighing about 40 carats. The provenance given for the latter states that they were sold by a dealer in 1838 to Russia's Princess Bagration, who resold them to a dealer.

Another intriguing reference to Marie Antoinette earrings is provided in a recent book, *Imperial Wedding of Old Paris*, by Nancy Becker. She describes large pear-shape diamond ear drops that originally belonged to Queen Marie Antoinette worn by Empress Eugenie when she was married in Notre-Dame Cathedral in 1853. They were a wedding gift from her husband-to-be, Napoleon III. The author references Madame Carette, Empress Eugenie's reader, who in her 1889 book *My Mistress the Empress Eugenie* recounted: "The personal jewels of the Empress consisted of a casket of the greatest value. Among others, there were some magnificent earrings, shaped like large pears, in diamonds, which originally belonged to the Queen Marie Antoinette, the Empress obtaining possession of them on her marriage, together with a necklace of most valuable pearls . . ." The empress Eugenie took her personal jewelry with her to England following the defeat of Napoleon III in the Franco-Prussian War. She sold most of her jewelry between 1870 and 1872.

So, are any of these Marie Antoinette earrings the ones purchased by Marjorie Merriweather Post from Cartier in London in 1928? Perhaps! When Cartier purchased the diamond earrings earlier that year from Prince Felix Felixovich Yusupov, the prince provided a written affidavit signed by his mother, Zinaida Nikolayevna, attesting to their authenticity. The document stated that the earrings had never been reset and that they had been in the possession of their family for a period exceeding one hundred years, and: "According to tradition, they were one of the last presents of Louis XVI to his queen; she wore them constantly; they were found in her pocket after the arrest of the French royal family at Varennes." She also adds that the earrings were acquired by her great-grandmother, Princess Tatiana Yusupova, in the first years of the nineteenth century, and that portraits of her ancestors wearing the jewels existed, but they remained in Russia with all documents concerning the family jewels. Prince Felix Yusupov,

LEFT
Portrait of Princess Tatiana Alexandrovna Yusupova, dated 1875, by French artist Jean-Baptist Marie Fouque, in which she is wearing the Marie Antoinette Diamond Earrings. (State Museum of the History of St. Petersburg)

OPPOSITE
The 20.34- and 14.25-carat Marie Antoinette diamonds removed from their settings to highlight their size and shape differences.

16

perhaps best known for his role in the assassination of Grigori Rasputin and for marrying the niece of Tsar Nicholas II, claimed to be the sole family heir and recounted how he was able to bring some of the family jewels abroad during the Russian Revolution.

Tatiana Vasilievna was a niece of the noted Russian military leader Grigori Potemkin and a lady-in-waiting for Empress Catherine the Great. She married Prince Nikolai Yusupov and successfully helped manage his estates into a sizable fortune. According to a 1905 biography by Grand Duke Nicholas Mikhailovich, she had a passion for precious stones and bought them all over Europe. Her wealth enabled her to assemble a magnificent collection that included the Polar Star Diamond, the La Pelegrina Pearl, and the earrings of Marie Antoinette. The collection was greatly expanded by her grandson (Prince Felix's grandfather), Prince Nikolai Yusupov (same name as his grandfather), and there is a portrait of his wife, Princess Tatiana Alexandrovna Yusupova, dated 1875, by French artist Jean-Baptist Marie Fouque in which she is wearing the Marie Antoinette earrings. Prince

Nikolai's mother's will, dated 1893, dictates that the jewels kept at her residence in St. Petersburg, including the Polar Star and her "earrings, so-called pendants of the Queen Marie Antoinette," go to her granddaughter Princess Zinaida Yusupova, mother of Prince Felix.

The earliest documented presence of the diamond earrings belonging to the Yusupov family is the 1875 portrait, leaving open the possibility that the Marie Antoinette earrings worn at Empress Eugenie's nuptials could have been purchased by the Yusupovs when the empress's jewels were sold in 1870–1872, or from the 1874 Geneva auction. Either option is inconsistent with Prince Felix's statement that the earrings have been in his family for over a hundred years, but there are no documents by which to evaluate the accuracy of that claim.

We have reviewed the testimony from historians, archives, and researchers, but what do the earrings themselves have to say about their original owner? The large pear-shape diamonds were removed from their settings at the Smithsonian Institution and their weights determined to be 20.34 and 14.25 carats. Spectroscopy measurements showed that both diamonds are the relatively rare type IIa stones, i.e., of highest purity, and are colorless to near colorless with good clarity. The quality and sizes of the diamonds are exceptional even by today's standards and would have been even more so in late 1700s Europe. Also, at that time India and Brazil were the only significant sources of diamonds, and the smaller supply limited ownership, especially of large diamonds, to the aristocracy. Finally, the cuts of the diamonds and the style of the settings are consistent with late eighteenth century French craftsmanship. The evidence, therefore, provided by the earrings strongly indicates that more than two hundred years ago they dangled from the ears of someone very wealthy and powerful. A French queen, perhaps?

Marie Louise Diadem

For the occasion of his marriage to Marie Louise of Austria, French Emperor Napoleon I presented her with a dazzling display of emerald and diamond jewelry that included a diadem, necklace, comb, belt buckle, and earrings. The court jewelers scoured Europe to locate sufficient diamonds and gems of exceptional quality, and amazingly completed the pieces in just a few months. Napoleon gave the jewels as gifts for Marie Louise's private collection, and as such they were never state property. The diadem is one of the most exquisite ever created and representative of the high level of jewelry craftsmanship in Paris at the time. The seventy-nine Colombian emeralds of exceptional quality—featuring a 12-carat central emerald and a blinding array of approximately one thousand old-mine-cut diamonds, with a total weight of more than 700 carats—were set in silver and gold in an elaborate design of scrolls, palmettes, and medallions. The original emeralds were removed by Van Cleef & Arpels in 1954–1956 and replaced with Persian turquoise.

During the period of turmoil following the French Revolution, a young General Napoleon Bonaparte waged a series of brilliant military campaigns that revitalized France as a European power, and in 1799 he assumed the position of first consul of the republic. His political acumen and popularity with the French people led him to crown himself emperor of France in 1804. By 1807, his patronage, and that of the French aristocracy, supported more than four hundred jewelers in Paris. When Napoleon married his second wife, Marie Louise of Austria, in a public ceremony in the spring of 1810, Honoré de Balzac wrote in *La Paix du Ménage* of the imperial wedding: "Diamonds glittered everywhere, so much so that it seemed as if the wealth of the whole world was concentrated in Paris. . . . Never had the diamond been so sought after, never had it cost so much." Napoleon commissioned court jeweler François-Régnault Nitot (of the House of Etienne Nitot et Fils) to create two spectacular jewelry parures (sets of matching jewelry), one with diamonds and opals and the second with emeralds and diamonds, as wedding gifts for his bride.

Following a series of disastrous military defeats, Napoleon abdicated in 1814 and was ultimately exiled to St. Helena. After his abdication, Marie Louise refused to accompany Napoleon into exile, and instead she and her son joined her family in Vienna. She returned the crown jewels but took with her the parures as her personal property.

After Marie Louise's death in Parma in 1847, the emerald and diamond parure, including the diadem, was bequeathed to her Habsburg aunt, Princess Elisabeth of Savoy. It was then passed to her son, Archduke Leopold (godson and cousin of Marie Louise), and then to his cousin Archduke Karl Albrecht. Archduke Albrecht and his wife were interned by the Germans during World War II, but after the war the family emigrated to Sweden, taking the family jewels with them.

LEFT
In January 1967, Van Cleef & Arpels loaned the Marie Louise Diadem to Marjorie Merriweather Post, accompanied by Colonel C. Michael Paul, to wear to the Red Cross Ball in Palm Beach, Florida.

OPPOSITE
The Marie Louise Diadem was donated to the Smithsonian National Gem Collection by Marjorie Merriweather Post in 1971. The original emeralds were removed by Van Cleef & Arpels in 1954–1956 and replaced with Persian turquoise. The magnificent antique diadem retains the original approximately one thousand diamonds, totaling about 700 carats, set in silver and gold.

This diamond and emerald parure was given by Napoleon I to Empress Marie Louise as a wedding gift in 1810. Van Cleef & Arpels acquired the diadem in 1952 and replaced the emeralds with turquoise. The emerald and diamond necklace and earrings are now on display in the Louvre.

Model wearing the Marie Louise Diadem with the original emeralds. (Erwin Blumenfeld for January 31, 1955, issue of *Life*.)

Following the archduke's death in 1951, his widow and son, Archduke Karl Stefan, attempted to sell pieces from the Marie Louise emerald and diamond parure, but the documentation for the jewels had been left behind in Poland, now behind the Iron Curtain, and could not easily be retrieved. Without proof of ownership, major jewelry firms were reluctant to complete a deal. The family's attorneys prepared an affidavit that was signed by the widow and her son, in which they attested that the set of jewels—emerald and diamond tiara, smaller comb tiara, ear clips, and belt clip—were among the personal jewels belonging to Empress Marie Louise, and that "according to the opinion of the imperial family of Austria, these jewels are the nicest ones among all the other jewels of Empress Marie Louise." They stated that through inheritance these jewels came to Archduke Karl Albrecht Nicolas Leo Gratien, husband and father of the undersigned, and that the jewels, until their departure for Sweden, were kept in a case in the shape of a saddle with documents of ownership. In 1952, the diadem and belt clip were purchased by Van Cleef & Arpels. The remaining pieces were retained by the family and eventually sold.

Van Cleef & Arpels displayed the newly acquired diadem in their New York store window on Fifth Avenue. Eventually bowing to overtures from enthusiastic clients, they agreed to remove the emeralds from the diadem and buckle and reset them into modern pieces. The emeralds were removed from the tiara during the period from May 1954 to June 1956 and sold individually in pieces of jewelry. In 1955, *Life* magazine proclaimed, "Napoleonic Tiara Is Torn Up" and showed the emeralds mounted in a newly designed platinum necklace, earrings, ring, and clip. Claude Arpels confirmed: "As each emerald was sold, we replaced it with a turquoise." Undoubtedly, turquoise was chosen as the replacement stone because it was relatively inexpensive and could easily be fashioned to fit the empty settings in the diadem. In 1962, the diadem with turquoise was displayed in the Louvre Museum in Paris along with the necklace, earrings, and comb, as part of a special exhibition on Empress Marie Louise. In 2004, the Louvre purchased the emerald and diamond necklace and ear clips that were part of the original Marie Louise parure. They are exhibited in the Galerie d'Apollon, where one can appreciate firsthand the exceptional quality and beauty of the emeralds so painstakingly sourced by Napoleon's court jewelers so as to be fitting for a gift from the emperor of France to his new bride. In 1999, one of the Van Cleef & Arpels brooches, set with nine emeralds from the diadem, sold at a Christie's auction for $178,500.

In an adroit marketing move, Van Cleef & Arpels loaned the now turquoise and diamond diadem to their special client Marjorie Merriweather Post to wear to the Red Cross Ball in Palm Beach, Florida, in January 1967. The diadem was the talk of the ball and clearly made an impression with Ms. Post. Claude Arpels wrote to her in 1971 indicating that there were several clients inquiring about the diadem, but because of her interest, he was offering it to her first, at a special price; she quickly agreed to purchase it for the Smithsonian gem collection. In April 1971, she donated the funds to the Smithsonian to purchase the diadem. It was exhibited in the Museum of History and Technology until 1973, when it was transferred to the growing National Gem Collection in the National Museum of Natural History.

Once it was at the Smithsonian, Ms. Post queried Van Cleef & Arpels about the possibility of replacing the turquoise with simulated emeralds to better show the original look of the diadem. They strongly discouraged the idea, explaining that the antique frame was too fragile to consider such modification. In fact, during the renovation of the Smithsonian gem and minerals galleries in the mid-1990s, the museum undertook a restoration of the historic diadem. The intricate framework had been soldered and repaired in several places as a result of its various journeys and modifications. An expert jewelry conservator, over several months, disassembled, repaired, and cleaned it, restoring it to its original glory—minus the emeralds, of course. The museum continues its quest to acquire at least some of the emeralds that were removed from the diadem, which, seen next to their original setting, would be a thrill for many visitors and curators.

Napoleon Diamond Necklace

The diamond necklace that Napoleon I lavished upon Empress Marie Louise after the birth of his long-awaited son and heir has survived intact and today wows Smithsonian visitors as it once did nineteenth-century admirers. It is a masterpiece of rare beauty and one of the great diamond necklaces of history. The necklace was fabricated by Paris jeweler Etienne Nitot et Fils in 1811. The design of the necklace departed from the ornate style of the period by highlighting the magnificent diamonds in a simple, elegant setting. The 234 diamonds set in silver and gold weigh about 263 carats. The largest diamond is about 10.4 carats, and as far as is known, none of the diamonds have ever been removed from the original setting. A recent study of the diamonds showed that they were all colorless, or near colorless, and of the very highest quality, even by today's standards. There are several portraits of Marie Louise wearing the necklace, both from her time in France and while residing in Parma.

There was much rejoicing when, a year after their marriage, Empress Marie Louise gave birth to Napoleon's son, the much-anticipated heir and the future king of Rome. The extravagance of his gift to the empress in celebration of the birth suggests the magnitude of Napoleon's joy—a magnificent necklace set with more than 260 carats of the finest quality diamonds. In order to appreciate the impact of this great necklace, then and now, it must be noted that in 1811, miners had not yet discovered South Africa's rich diamond fields. The major sources of diamonds at the time were India and Brazil, and production was such that only the rich and influential owned diamonds. Of course, this is exactly why Napoleon gave such a necklace as a demonstration to the world of his unparalleled wealth and power. An appraisal of the necklace at the time by jeweler Ernst Palscho, apparently to verify the value of the completed necklace, is preserved in the French National Archives and itemizes the weight and cost of each of the necklace's 234 diamonds and gives a total value of 376,275 gold francs. It is signed as received by Chamberlain, Count of Montesquin on July 13, 1811, with payment authorized on July 18. This was an enormous sum of money at that time, approximately equivalent to the empress's yearly household budget.

After Napoleon's abdication in 1814, Marie Louise retained the diamond necklace while serving as ruler of the duchies of Parma, Piacenza, and Guastalla. She bequeathed the diamond necklace to her sister-in-law Archduchess Sophie, wife of Archduke Charles Francis Joseph of the House of Habsburg-Lorraine. Sophie removed two of the round brilliant diamonds to shorten the necklace. After her death in 1872, the necklace eventually passed to her third son, Archduke Charles Louis. In 1929, his wife, the Archduchess Maria Theresa, described the sensation created when the necklace was worn to the coronation of Tsar Alexander III in 1883: "The imperial family as well as the grand dames of Russia desired so much to see it closely that, in order to content them, it had to be exhibited at certain fixed hours under guard." Archduke Charles Louis left the necklace to his three daughters. However, like many royal families in the aftermath of World War I, they were forced to sell the family jewels for needed income.

LEFT
This painting by Giovanni Battista Borghesi, dated 1837–1839, shows Marie Louise wearing her diamond necklace while she was Duchess of Parma. (Galleria Nazionale di Parma)

OPPOSITE
The diamond necklace given by Napoleon I to Empress Marie Louise to celebrate the birth of their son in 1811. The 234 diamonds set in silver and gold weigh about 263 carats. The largest diamond is about 10.4 carats.

ABOVE
Detail from the necklace showing
the mountings for the briolette and
pendeloque (pear-shape) diamonds.

LEFT
The original leather case made for
the necklace in Paris was donated
with the necklace and is part of the
Smithsonian National Gem Collection.
It is in the empress's official colors and
inscribed with her initials.

OPPOSITE
Photograph of Archduchess Maria
Theresa wearing the Napoleon
Diamond Necklace, circa 1900.

The failed attempts to sell the necklace—first to the king of Egypt, then to a buyer in Amsterdam—led to one of the more unusual episodes in the necklace's history. Apparently trusting the recommendation of a friend of a friend, in 1929, Archduchess Maria Theresa consigned the necklace for sale in America to people she had never met: "Colonel" Townsend, allegedly having served in the British military, and his wife, Princess Baronti, a woman who turned out to be neither his wife nor a princess. A complicated series of events unfolded, in which the Townsends enlisted the assistance of Archduke Leopold of Habsburg, Maria Theresa's grandnephew, self-described as one of the "poor Habsburgs." The Townsends negotiated multiple deals to sell the necklace, first to New York jeweler Harry Winston, then to attorney Harry Berenson, and later to Marjorie Merriweather Post (who eventually *did* purchase it for the Smithsonian in 1962). David Michel, a New York diamond dealer, finally bought it for $60,000 (considerably less than the original asking price of $450,000), intending to remove

and recut the diamonds. The Townsends sent $7,270 to Maria Theresa and kept the balance of $52,730 to cover their "expenses related to the sale," which included $20,000 for Leopold, $9,000 to rescind the Berenson deal, and $2,500 to a broker who helped locate potential buyers.

Not surprisingly, the archduchess (who had rescinded her contract with the Townsends long before they sold the necklace) was outraged. An emissary was dispatched from Austria to retrieve the necklace. After some weeks, they negotiated an arrangement with David Michel by which the archduchess paid him $50,000 to return the necklace and Michel agreed to absorb a $10,000 loss. An investigation by the district attorney's office made the necklace affair public, and the Napoleon necklace scandal was a front-page story on March 1, 1930. Grand larceny charges were handed down against the Townsends and Leopold, but by that time, the mysterious couple had fled and was never arrested. Only the hapless archduke was left to face trial. The combination of the Napoleon necklace and a nobleman defendant provided a spectacle that fascinated the press and the public during the summer, when Archduke Leopold chose to spend time in jail, in the "Tombs," over bail, and through the November trial. Ultimately, Leopold was acquitted of the charges.

The diamond necklace remained in the United States until it was returned to the Habsburg family in mid-1932. And what of the Townsends? The enigmatic "Mrs. Townsend," referred to by the American press as Princess Baronti, later using the name Gerveé Baronti, relates in her 1935 autobiography that following her quick exit from New York after the sale of the necklace, she traveled to Chicago and then to India. She was, in fact, never married to Mr. Townsend (he already had a wife) and mentions that she saw him for the last time in England, as he was making plans to move to Japan. Gerveé Baronti died in 1960 in a state hospital in Florida. Following the public drama of the necklace affair, Archduke Leopold slipped into obscurity, eventually working as a janitor; he died in Connecticut in 1958.

The diamond necklace finally left the Habsburg family in 1948 when it was sold by Archduchess Maria Theresa's son-in-law, Prince Francis Joseph II of Liechtenstein, in agreement with his cousin Archduke Max Eugene, to Paul L. Weiller, a gem dealer in Geneva and Paris. Harry Winston acquired the necklace in 1960 and sold it to Marjorie Merriweather Post in 1962, who later that year gifted it to the Smithsonian Institution.

Blue Heart Diamond

In June 1964, Marjorie Merriweather Post arrived at the office of Smithsonian secretary Leonard Carmichael bearing gifts, her bag brimming with jewelry that she was gifting to the Smithsonian Institution. Among the treasures was a magnificent blue diamond that she had purchased from Harry Winston only two months earlier. The 30.62-carat diamond was cut into a heart shape and set into a ring surrounded by twenty-five round brilliant, colorless diamonds. With its large size, deep-blue color, and exquisite cut, the Blue Heart Diamond was, and is, one of the world's great diamonds and considered by many aficionados to be the most beautiful of all blue diamonds. Today it is a visitor favorite in the Gem Collection gallery, appreciated for its singular beauty and glamorous past.

Blue diamonds are exceedingly rare, accounting for fewer than one in two hundred thousand mined diamonds, and the Premier mine (now Cullinan mine) in South Africa is a primary source of these rare stones. According to the De Beers archives, in October 1909, an exceptional 100.5-carat deep-blue diamond was found at the Premier mine, and in December of that same year, the rough crystal was sold for £3,979.16 (equivalent to approximately $600,000 in 2016) to diamond cutter Atanik Eknayan of Neuilly, near Paris. He had the stone faceted into a 30.62-carat heart-shape brilliant and later in 1910 sold it to jeweler Pierre Cartier. Cartier used the blue diamond as the centerpiece in a corsage ornament of lily-of-the-valley design, accented with a pink and a smaller blue diamond. He sold it in a necklace in 1911 to heiress and philanthropist María Unzué de Alvear (1862–1950) of Buenos Aires, Argentina, and in 1936 she gave it as a wedding present to her niece, Angela Gonzalez Alzaga. Because of this part of its history, the diamond is sometimes referred to as the Unzué Heart.

Van Cleef & Arpels of Vendôme, France, purchased the necklace in 1953. According to a member of the Unzué family, the price exceeded that of a very fine house in Buenos Aires at the time. Van Cleef & Arpels reset the colored diamonds into a pendant that was worn in June of that year by actress and ballerina Zizi Jeanmaire to a charity ball in the Orangerie at Versailles. Later in 1953, the pendant was purchased by Swiss industrialist Baron

LEFT
The unset Blue Heart Diamond measures 20.15×19.84×11.85 millimeters (0.8×0.78×0.47 inches); shown with the Hope Diamond; the fancy deep-blue color of both diamonds originates from trace amounts of the element boron that were incorporated into the diamond structures as they formed more than a billion years ago hundreds of miles below the earth's surface.

OPPOSITE
The 30.62-carat Blue Heart Diamond, surrounded by twenty-five round brilliant diamonds, totaling 1.63 carats, in a ring setting designed by Harry Winston Inc.

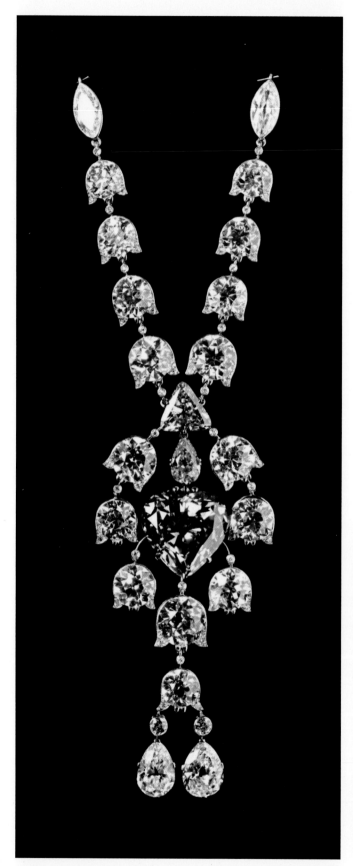

Hans Heinrich Thyssen-Bornemisza. It was one of many extravagant gifts, including an island in Jamaica, that he showered upon fashion model Nina Sheila Dyer, with whom he had a longtime affair, beginning when she was seventeen years old. They were finally married in 1954 but divorced ten months later, after she had an affair. Soon after, Ms. Dyer married Prince Sadruddin Aga Khan, and in 1959 she sold the Blue Heart pendant to New York jeweler Harry Winston.

For Mr. Winston, this was an opportunity finally realized. He had first become aware of the Blue Heart Diamond during a trip to Buenos Aries in 1946, but he was unsuccessful in his attempt to acquire the diamond at that time. Winston reset the Blue Heart into its current ring setting and sold it to Marjorie Merriweather Post in April 1964. Two months later, she gifted the diamond to the Smithsonian Institution, reserving the right to "use, enjoy, and take into her possession . . . during her natural life for such period or periods of time as she shall desire." The diamond was loaned back to her in 1964, 1968, and 1970. From June through August 1971, the Blue Heart was featured in the Kimberley Centennial Exhibition in South Africa that celebrated one hundred years of diamond mining in that country. At the Smithsonian Institution's National Museum of Natural History, the Blue Heart is admired by several million visitors annually. Once, when walking through the museum's gem gallery, Ms. Post was overheard saying to her guests that "my blue diamond is prettier than that other one"—referring, of course, to the Hope Diamond. Many would agree with her.

OPPOSITE, LEFT
Lily-of-the-valley corsage ornament fabricated by Cartier in 1910, featuring the Blue Heart Diamond, along with a 2.05-carat pear-shape pink diamond and triangular blue diamond weighing 3.81 carats.

OPPOSITE, RIGHT
María Unzué de Alvear (1862–1950) of Buenos Aires, Argentina, with her niece, Angela Alzaga Unzué, purchased the necklace with the Blue Heart diamond from Pierre Cartier in 1911.

ABOVE
Nina Dyer in 1954 with the Van Cleef & Arpels necklace featuring the Blue Heart Diamond.

RIGHT
The Blue Heart pendant designed by Van Cleef & Arpels in 1953, also including the pink and blue diamonds from the original corsage ornament.

Post Emerald Necklace

Marjorie Merriweather Post purchased the impressive art deco necklace of more than 650 carats (over a quarter pound) of Indian-cut emeralds set with platinum and approximately 1,350 diamonds in 1929, and she donated it to the Smithsonian Institution in 1964. The magnificent necklace was fabricated by Cartier in London in 1928 and finished in New York in 1929.

ABOVE
Maharaja Sir Vijaysinhji ruled the western Indian state of Rajpipla from 1915 until its merger with the Union of India in 1948. Portrait by Vandyk, London, circa 1922.

In 1876, Queen Victoria was proclaimed empress of India, and in the following decades, jewelry featuring Indian motifs and Mogul-cut gems became all the rage in England and France. By the early twentieth century, the House of Cartier actively sought Indian stones, particularly emeralds and diamonds, which they set in platinum jewelry inspired by Persian and Indian patterns. Conversely, Indian rulers were fascinated by European jewelry, and they sent their family jewels to Cartier and others for reworking into newly fashionable designs set with platinum and diamonds. In 1925, the maharaja of Patiala commissioned Cartier to remodel his extraordinary collection of crown jewels, which led to several years of work and inspired other maharajas to do the same. This cross-cultural exchange of jewels and traditions propelled the bold and innovative art deco designs that characterized much of Cartier's output in the 1920s and 1930s.

Although emeralds were the gems most prized from India, in fact, the original source for these deep-green stones was Colombia. Prior to the sixteenth century, emeralds generally were not major players in the gem world. Small mines in Egypt and Austria provided only a modest number of unspectacular green stones. It all changed in 1537 when the Spanish discovered the rich emerald sources in the Colombian jungles near the sites of the present-day Chivor and Muzo mines. For more than one hundred years, they shipped huge quantities of large, deep-green crystals to Europe and Asia. This sudden influx of high-quality emeralds resulted in a gradual acceptance of this exotic gem by European jewelers and their clients, but they were immediately and enthusiastically embraced by the courts of India and Persia. In addition to the gems' impressive sizes and rich hues, the Moguls and those in other Muslim regions prized them because green is Islam's sacred color. Consequently, the great bulk of these New World gems were fashioned into extravagant necklaces, turban ornaments, brooches, and all forms of adornments by the jewelers' workshops of the Indian maharajas. Emeralds' popularity quickly established them as elite gems that were valued equally with pearls and rubies. Indian cutters typically rounded and polished, rather than faceted, the prism-shape crystals to preserve size and weight.

OPPOSITE
The pavé-set diamond chain suspends twenty-three graduated baroque cabochon emerald drops, the largest more than two centimeters in diameter and weighing about 25 carats, each surmounted by a smaller emerald bead; an additional emerald drop hangs from the ornate diamond and platinum clasp.

One of Cartier's very good customers was Marjorie Merriweather Post, and she purchased the emerald necklace from them in 1929. Unfortunately, aside from that, Cartier is not able to provide a detailed history of this necklace, as their London branch records were destroyed during the Blitz of World War II. They can confirm that the emeralds are of Indian origin, referring to where the gems were fashioned, not mined. The rounded shapes of the emeralds, and the fact that they were drilled so that they could be attached to the necklace with rods, are consistent with Mogul craftsmanship, suggesting that the gems had been fashioned in India. The style of the necklace is similar to those traditionally worn by Indian male royalty for ceremonial occasions. The Hillwood Estate, Museum & Gardens' archives suggest that the original necklace likely belonged to the maharaja of Rajpipla. If so, the necklace was almost certainly remodeled by Cartier before selling it to Ms. Post. Along with the necklace, Ms. Post also purchased a stunning emerald pendant brooch that featured a large center stone with a Mogul inscription. She wore it as a pendant attached to the emerald necklace and later had it altered so that it could also be worn as a brooch.

OPPOSITE
Marjorie Merriweather Post wearing the Post Emerald Necklace. Portrait by Frank O. Salisbury, 1953.

RIGHT
The emerald beads are drilled and attached to the necklace with metal rods, typical of the Mogul style.

ABOVE RIGHT
At the 1929 Palm Beach Everglades Ball, Ms. Post introduced her new emerald necklace with attached pendant, which must have made a colorful pairing with her bright orange Juliet costume.

Maximilian Emerald

The Maximilian Emerald weighs 21.04 carats and measures 20.02×14.7×9.6 millimeters (0.8×0.6×0.4 inches). It has a rich green color and classic emerald cut. The gem's interior shows wispy inclusions and veils that are typical of Colombian stones.

Was this 21.04-carat emerald once the property of Emperor Maximilian of Mexico? Perhaps, but very likely we will never know for sure. What we do know is that when Marjorie Merriweather Post purchased the emerald in 1928, likely from Cartier, she was told the gem had once belonged to the ill-fated Mexican monarch.

When not occupied with official duties, Ferdinand Maximilian pursued an active interest in natural history, particularly botany. In 1860, he traveled to the tropical forests of the Amazon basin to study their famed plant diversity. On that trip, he purchased two large diamond crystals, one that was later cut into a 42-carat gem that he wore in a small bag around his neck at his execution, and another that was fashioned into a 33-carat stone that he presented to his wife, Carlota. Some believe he might have acquired the emerald on this same expedition—intriguing idea, but only speculation.

Ferdinand Maximilian Joseph, archduke of Austria, was the brother of Austrian Emperor Franz Joseph of the House of Habsburg. In 1856, at the age of twenty-four, he married Princess Carlota, daughter of Emperor Leopold I of Belgium. Following his retirement from the Austrian Navy, he was enticed by Napoleon III and a group of exiled Mexican landowners living in Europe to be appointed emperor of Mexico. Napoleon III had imperialistic ambitions in North America, while the exiles, and the Catholic Church, hoped to regain property confiscated by the Mexican government of Benito Juárez. Maximilian was oblivious to the political intrigue and financial motives and was convinced by all involved that he would be warmly welcomed by the Mexican people as a replacement for Juárez. The French army, with early support from Britain and Spain, occupied Mexico City in 1863 and drove Juárez and his army almost to the Texas border by April 1865. Ferdinand Maximilian and his wife Carlota were crowned emperor and empress of Mexico on June 10, 1864, in Mexico City. He quickly antagonized his European backers when he upheld the land reform laws of Juárez and advocated for educating the poor. The situation went from bad to worse for Maximilian when Napoleon III realized he had underestimated the cost of his Mexican foray and the strength of the Mexican resistance, and the United States, no longer preoccupied by the Civil War, demanded in 1865 that France remove its troops from Mexico. By March 1867, the last French soldiers left Mexico, and Juárez and his army quickly reoccupied Mexico City. The duty-bound Emperor Maximilian refused to leave the Mexican people, and he was arrested by Juárez. Empress Carlota traveled to Europe to beseech Napoleon III, and then the pope, to aid her husband, but they refused. Emperor Maximilian, age thirty-five, was executed by firing squad on June 10, 1867, on a hill outside of Querétaro. Carlota's failure to save her beloved husband sent her into a deep depression and state of mental illness that she endured until her death in 1927.

In 1929, shortly after purchasing the emerald, Ms. Post wore it in a ring for her presentation at the Court of St. James. Her granddaughter, Ellen Charles, recounted that Ms. Post later wore the emerald ring during the coronation of King George VI in 1937, and after attending an event at Buckingham Palace, she realized that the emerald was missing from the setting. Fortunately, a phone call to the palace resulted in the quick recovery of the gem.

In 1949, Cartier reset the emerald into its current platinum setting, flanked on either side by three baguette diamonds. Ms. Post gifted the ring to the Smithsonian Institution in 1964.

Post Diamond Tiara

The early nineteenth-century Post Diamond Tiara is believed to have been made in France. A tiara is a form of crown that sits at the front of the head, typically a semicircular band of metal set with diamonds and other gems. The garland of wild roses design displays flower petals and leaves pavé-set with 1,198 old-mine-cut and rose-cut diamonds, and flower centers set with larger rose-cut diamonds; the large front flower head is set with a brilliant-cut diamond at its center. The flower motifs are attached *en tremblant*—that is, mounted on trembler springs, so that every movement by the wearer would enhance the brilliance and sparkle of the diamonds. The tiara is made of silver and gold and is accompanied by a pair of matching floral sprays that may be worn as brooches or to extend the length of the tiara. An antique tiara such as this was fabricated by hand, with each diamond set into holes cut specifically for that stone, a process that took thousands of hours. The Post Diamond Tiara, once the property of British noble Lord Paul Ayshford Methuen, was purchased at a Sotheby's auction in London for the Smithsonian by Marjorie Merriweather Post in 1970.

Jadeite Dragon Vase

Smithsonian curator George Switzer saw this spectacular jadeite vase while visiting T. C. Chang in Santa Monica, California, in 1968. He described it as the "finest piece he had ever seen in his entire life" and declared the asking price of $80,000 "a bargain." When the Smithsonian approached patron Marjorie Merriweather Post about sponsoring the purchase, her prompt reply was, "OK, buy." She transferred 250 shares of IBM stock to the Smithsonian, and the deal was completed in December 1968.

The extraordinary vase is carved from a type of jade: The name "jade" is applied to two different minerals that have similar physical properties, nephrite and jadeite. Nephrite ranges from creamy white to green to almost black in color. Jadeite is white, green, or rarely, purple, and sometimes all three colors appear within a single piece, as seen in the Dragon Vase. The internal arrangements of jadeite and nephrite crystals make jade tough, allowing it to be useful historically for making weapons and tools, as well as delicate carvings. Jadeite is also valued for its color: Traces of chromium impart the brilliant green; manganese produces the purple, mauve, and lilac shades; and the white parts are impurity-free.

Historically, most jade used in China was nephrite, but in the late 1700s, China started to import large quantities of high-quality jadeite from Burma (also known as Myanmar).

OPPOSITE
The antique diamond tiara with matching floral sprays was likely fabricated in France in the early nineteenth century.

RIGHT
The Jadeite Dragon Vase stands 50 centimeters (19.7 inches) tall and is carved from a single piece of rare multicolored jadeite from Burma. The Chinese carving is in the Ch'ien-Lung (Qianlong) style, with his seal carved on the base, and was probably done in the early 1960s.

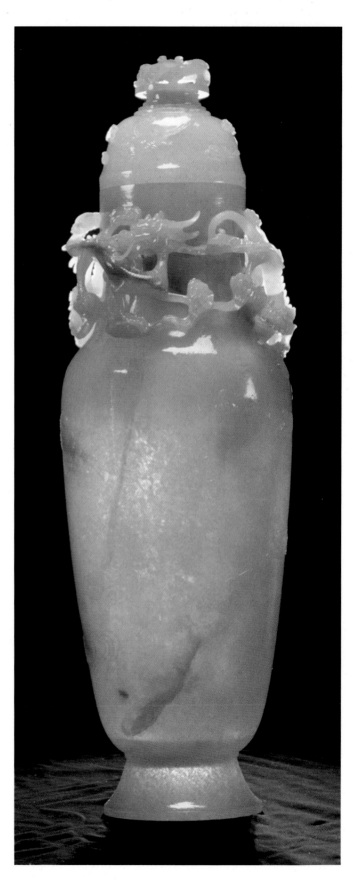

Logan Sapphire

The Logan Sapphire came into Rebecca "Polly" Pollard Guggenheim's possession in 1952 as a surprise gift from her husband, Colonel M. Robert Guggenheim. The extraordinary egg-size gem, one of the largest and finest known sapphires in the world, was originally mined from the gem-rich gravel deposits of Sri Lanka. The sapphire is a slightly violet-blue cushion mixed cut, measuring 49.23×38.26×20.56 millimeters (1.9×1.5×0.8 inches, about the size of a large chicken egg). It is an extraordinarily clean stone for its size, showing some of the "silk" (aka rutile inclusions) characteristic of Sri Lankan sapphires. At 422.98 carats (about 3 ounces), it is believed to be the second largest faceted blue sapphire gem known. (The largest reported faceted sapphire, 478 carats, was set by Cartier in the early 1920s for Queen Marie of Romania. It later was passed to the Greek royal family, and a 1964 news story reported it to be 460 carats.) It is not known when, or by whom, the stone was put into the current setting. In a 1958 article in *Ladies' Home Journal*, however, Polly states that the sapphire was "reset surrounded by twenty round-cut diamonds," suggesting that it might have been set after she acquired the stone. A close examination of the setting reveals that the round diamonds surrounding the sapphire likely were part of an antique necklace or bracelet that was repurposed to make the sapphire brooch/pendant.

OPPOSITE
The Logan Sapphire, weighing 422.98 carats and measuring 49.23 millimeters (almost 2 inches) tall, is set in a silver and gold brooch/pendant surrounded by twenty white brilliant-cut diamonds, having an estimated total weight of 16 carats.

When Polly Guggenheim Logan was once asked if her extraordinary 422.98-carat sapphire brooch was perhaps a bit large to wear, she simply responded: "I thought so too until I owned it." The gem was a surprise Christmas/anniversary gift from her second husband, Colonel M. Robert Guggenheim, in 1952. Seeing it among the silver wrapping paper and ribbon, she said in *Ladies' Home Journal* in 1958: "I was simply overcome. I had never seen it and thought it was very beautiful, but it had never occurred to me to think of it as something for me. It's simply not a stone you could wear casually. All I could think was, I'll never be able to wear it. And, of course, I loved it. How could I help loving something so beautiful?"

Rebecca "Polly" Pollard was born near Norfolk, Virginia, where her father was one of the leading developers of Virginia Beach and her mother was a Colonial Dame. After attending school in Staunton, Virginia, and finishing school in New York, she completed her studies in the arts at the School of the Museum of Fine Arts in Boston. It was in Boston that she caught the eye of Colonel Guggenheim. The beginning of their courtship coincided with her divorce in 1937, and the following year, vivacious Polly and the colonel were wed (his fourth marriage) in Miami aboard his 175-foot yacht, the *Firenze*.

Polly and the colonel lived on the *Firenze* for several years, and in 1942 settled in Washington, DC, living on the yacht moored in the Potomac River. They eventually purchased a twenty-acre city estate in northwest DC and built a fifty-nine-room Tudor-style mansion, which they called Firenze.

The Guggenheims entertained the most powerful players in Washington and were known for hosting elaborate parties at their Firenze house. Polly wore the sapphire frequently, to inaugural balls, White House dinners, embassy parties, and their many private dinners. *Ladies' Home Journal* in 1958 describes: "And the gigantic sapphire owned by Mrs. Robert Guggenheim of Washington, DC—424 carats [sic], the largest sapphire in the world. It is so heavy she usually wears it as a clip and then only on a gown with sleeves and strongly reinforced shoulder strap. Never on a strapless gown." Owning a world-renowned gem, Polly cautioned, is not always easy: "You can never forget about it. You have to have your clothes designed for it. People always notice when you wear it and mention it when you don't. And, really, sometimes I like to wear something else." She added, "I have to be careful; I don't dare drop it."

Polly became one of the leading figures of Washington society and was an active patron of the arts and a philanthropist. During

his military service, Colonel Guggenheim had befriended then-Lieutenant Dwight Eisenhower. They renewed their friendship in DC, and Polly became one of Ike's most successful campaign fundraisers. Although long having a reputation as a philanderer, Colonel Guggenheim's indiscretions became more public as time went on. Colonel Guggenheim died in 1959 after dining with his mistress in a Georgetown restaurant. Polly was ill in a hospital bed at the time. When the news was given to Polly, she sat up, feeling much recovered, asked for her dresses, and wanted out of the hospital.

A year later, in December 1960, Polly agreed to give her sapphire to the Smithsonian Institution; it was the first major gem added to the collection since the gift of the Hope Diamond from Harry Winston in 1958. Unlike the curse stories associated with the blue diamond, Polly considered her sapphire to be a good-luck stone. The agreement was structured such that she gave four sevenths of the gem immediately, and the remainder was donated in December 1961. In a letter to Smithsonian secretary Leonard Carmichael in October 1961, Polly expresses her wishes, referring to the sapphire gem: "which to the best of my knowledge is the largest such gem in existence in the world today. In view of the gem's unique nature, I should be greatly pleased if its use can be reserved for the wife of the president of the United States at such state and other occasions as may be appropriate." In fact, the sapphire has never been loaned for such a purpose.

Despite the completion of the gift in 1961, it would be almost a decade before the public would have a chance to see this great gem on exhibition. The gift agreement permitted the stone to remain with Polly throughout her life, or until she decided to return it to the museum.

In 1962, Polly married John Logan, a successful management consultant in Washington, DC. It was by all accounts a happy union, and Polly relished her role as hostess and philanthropist. She hosted a large birthday bash for Lyndon Johnson, and her annual Christmas parties at Firenze were among the highly anticipated social events of the year.

Polly wore the sapphire occasionally between the time of the gift and when she actually delivered it to the Smithsonian. At one family gathering, her grandson Douglas asked, "What is that big rock you are wearing?" Polly replied: "Come over here and thump it. It once belonged to a maharaja, and soon it will be in a museum." In April 1971, Polly sent the great sapphire brooch to the Smithsonian, and on June 22 of that year she formally presented it to Smithsonian curator Paul Desautels for public exhibition—where it remains today. According to the curator, when he asked Polly why she finally decided to give the famous sapphire to the museum, she replied: "Whenever I wore it, I was reminded of my cheating previous husband." Problem solved! But why hold on to the gem as long as she did? "It must be," she said, "the way anyone comes to feel about a possession that makes him feel good and happy. The longer you have it, the more you like it. But most things aren't as beautiful as jewels. Nothing gives a woman so much beauty. And most things you know you'll have to give up someday, no matter how much you love them. Your jewels, never!"

Polly Logan wearing her signature sapphire brooch, in an image published in the *Washington Evening Star*.

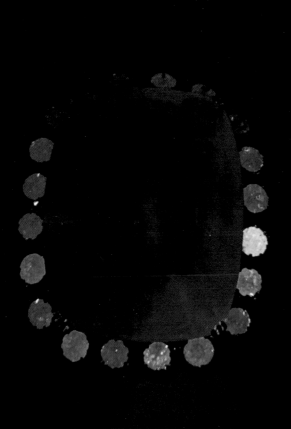

Mysterious Past

But what of the great sapphire itself and its history before showing up under Polly Guggenheim's Christmas tree? The Gemological Institute of America verified that the sapphire originated from Sri Lanka.

One might think that tracing the history of one of the world's largest sapphires should be reasonably straightforward, but there is still some mystery. Polly often related that the sapphire had once been owned by a maharaja and had been worn in a turban. A 1960 news story in the *Washington Star* announcing the donation of the sapphire to the Smithsonian recounts information from the jeweler who sold the stone to Colonel Guggenheim. He admits that the history is hard to authenticate (and almost certainly not true) that the first owner was a native who found the stone and was beheaded for failing to turn it in to his leader. It was then sold to a maharaja, and in the early 1900s purchased by an Englishman. It was shown at the New York World's Fair in 1939 and then remained in New York with the jeweler until it was purchased by Guggenheim. An article in 1941 in the *Ohio Jewish Chronicle* describes a 425-carat sapphire to be raffled by Sir Ellice Victor Sassoon, 3rd Baronet of Bombay, of the famous Anglo-American family, for the British war relief fund. He supposedly had acquired the gem from an Indian maharaja, and presumably was the Englishman referred to in the news story. The auction never took place.

Bismarck Sapphire Necklace

The spectacular 98.57-carat Bismarck Sapphire is exceptional for its outstanding clarity and rich blue color. Laboratory studies confirm its Burmese origin, and as such it is one of the largest and finest of its kind known. It measures 27.8×21.25×15.2 millimeters (1.1×0.84×0.6 inches). Correspondence at the time of the gift indicate that the sapphire was acquired by the Countess von Bismarck in India, likely during her 1926 honeymoon cruise. Cartier records indicate that the sapphire was originally set horizontally into a necklace in 1927 and reset as a bracelet in 1945. It was placed into the current setting in 1959 by Cartier in Paris, using one of their classic art deco designs.

OPPOSITE
In addition to the 98.6-carat Bismarck Sapphire, the Cartier necklace features eight small sapphires, 226 brilliant-cut diamonds, and eighty-six emerald-cut diamonds.

Countess Mona von Bismarck defined the art of living well. She followed the tried-and-true path to fame and fortune, and to accumulating a great jewelry collection, including the Bismarck Sapphire—she married early, she married often, and she married well. Frances "Mona" Strader, born in 1897, was the daughter of the owner of the Forkland horse farm near Lexington, Kentucky. In 1917, she married Henry James Schlesinger, a business associate of her father's who was eighteen years her senior and scion of the wealthiest family in Wisconsin. She divorced Schlesinger in 1920, and the following year married James Irving Bush, a bond salesman who was said to be the "best-looking man in America." She was twenty-four, and he was thirty-eight. The couple, especially Mona, enthusiastically embraced New York society, but Mona also spent time in Paris, and it was there in 1924 that she obtained her divorce from Bush. Mona's social and financial trajectory took a dramatic upturn in 1926 when she married Harrison Williams, twenty-four years her senior and the richest man in America, whose stated ambition was to become the country's first billionaire. Their honeymoon was a yearlong trip on his yacht *Warrior*, the largest in the world. It is during this journey at a stop in India that Mona likely acquired the extraordinary sapphire now in the Smithsonian collection.

Mr. and Mrs. Williams lived a lifestyle that has been described as social, sophisticated, and materially glamorous. *Time* magazine in 1929 suggested that "the only reason that the Harrison Williams don't live like princes is that princes can't afford to live like the Harrison Williams." They eventually owned five homes on two continents, including the fabulous Il Fortino on the Isle of Capri. The latter residence was built by French composer Georges Bizet on the foundation of the former palace of Roman emperor Tiberius and was noted for its fabulous gardens and sweeping views of the Bay of Naples. It was here that the "Queen of Capri" held court in the 1930s, and after World War II, entertained European nobility, artists and writers, and such notables as Benito Mussolini, the Kennedys, and Winston Churchill.

Mona was the showpiece of balls and galas and a regular feature in society pages; her more than fifty appearances in *Vogue* established her as a celebrity and fashion icon. She could afford the latest fashions from the finest design houses, and she became a favorite subject of photographer Cecil Beaton, who memorialized her great beauty and mesmerizing aquamarine eyes. In 1933, a group of leading Paris designers, including Mainbocher, Balenciaga, Molyneux, Lanvin, Lelong, and Chanel, declared Mona

44

the best-dressed woman in the world, the first ever such designation. She received the recognition again in 1934 and appeared on the list of best-dressed women nine more times between 1941 and 1957; in 1958, she was among the first group inducted into the Fashion Hall of Fame. Mona's celebrity was such that she was referred to in two Cole Porter musicals and in the opening scene of the movie *Mrs. Skeffington*, in which Bette Davis reassures herself by remarking: "You're Venus and Mrs. Harrison Williams combined; you're just too beautiful to live." She is also the subject of a hauntingly beautiful portrait painted by her friend Salvador Dalí.

Harrison Williams's fortune was severely diminished by the 1929 stock market crash, thereby ending his dream of being a billionaire and forcing a retrenchment in their lavish lifestyle. They eventually closed four of their five homes. Harrison Williams died in 1953, and that same year Mona became a countess with her marriage to Count Edward von Bismarck, grandson of Germany's "Iron Chancellor," Otto von Bismarck. Eddie was a longtime mentor to Mona on intricacies of European society, and a close friend. The couple lived happily and relatively quietly in apartments in Rome and Paris until Eddie's death in 1970. Later that year, Mona married Eddie's physician, Umberto de Martini. Only after his death in a car crash in 1979 did Mona realize that during the marriage he had lied, cheated, and stolen from her. She quickly dropped the Martini name and remained the Countess von Bismarck until she died in Paris in 1983.

In May 1967, during a visit by Mr. Cott from the National Gallery of Art to the Countess von Bismarck in Capri to receive a donated painting, she mentioned she had a very valuable antique diamond and sapphire necklace that she was contemplating giving to a museum. Mr. Cott suggested that the Smithsonian might be interested. Her financial manager contacted Smithsonian secretary S. Dillon Ripley, and arrangements were agreed upon to transport the necklace from Paris to Washington, DC. The gift was formally accepted into the Smithsonian National Gem Collection later that year.

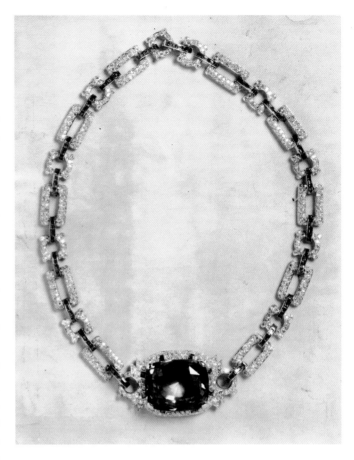

ABOVE
The Bismarck Sapphire in the original Cartier necklace (1927–1945).

OPPOSITE
Mona Williams in *Vogue* in 1933 wearing a dress by Chanel with pearls.

Rosser Reeves Star Ruby

With its translucent, rich red color and a sharp, well-centered star, the Rosser Reeves Star Ruby might be the largest and finest in the world. It was mined in Sri Lanka sometime before 1953. Years before the ruby was acquired by Mr. Reeves, Smithsonian gem curator George Switzer was on the trail of this magnificent ruby as the next prize (following the arrival of the Hope Diamond) for the growing National Gem Collection. The oval cabochon-cut stone measures 31.5×26.5×19.7 millimeters (1.25×1.04×0.78 inches) and is 138.72 carats. (It weighed 140 carats originally, but was recut and polished to its current weight to better center the star and remove surface scratches.) It was donated to the Smithsonian Institution in 1965 by legendary adman ("Prince of Hard Sell") Rosser Reeves II. In addition to penning memorable marketing slogans (Wonder Bread: "Helps build strong bodies 12 ways"), he wrote the book *Reality in Advertising*, which forever changed that business. He was the first person to apply marketing and advertising strategies on the new medium of television to a presidential campaign. He wrote all of the television spots for Dwight Eisenhower's election in 1952. According to the *New York Times* (October 21, 1979), when asked about his net worth, Rosser Reeves responded: "You don't give away a million-dollar ruby if you're impoverished."

In September 1961, Switzer received a call from Edwin I. Firestone, of the Boston jewelry firm Firestone and Parson. He described a rare, large star ruby of more than 100 carats. Firestone was representing the owner of the stone, Paul Fisher, whose father, Robert, had purchased the Sri Lankan ruby, in a simple antique setting, at an auction in London in 1953. Nothing was known about the earlier history of the stone. Switzer responded by letter, deeply regretting that "the museum is not in a position to be able to negotiate for the gem".

Enter adman Rosser Reeves II; son of a minister from Richmond, Virginia, he helped found the Ted Bates advertising agency in 1940.

In the summer of 1965, when Switzer visited Robert Nelson and Paul Fisher in New York, he was told that Rosser Reeves was interested in purchasing the star ruby. Reeves wrote in a letter to Switzer, "You know with what passion I collect and love jewels," and added, "I have not heard whether I own the 138-carat star ruby; the minute I get some definitive word, I will let you know." Switzer responded with classic curatorial chutzpah: "There is nothing I would rather see than that ruby in the Smithsonian collection with your name on it." Switzer's hope (strategy) at that point was that Reeves would acquire the ruby and loan it for exhibition at the Smithsonian, and then, of course, at some time donate it. But an afternoon meeting at New York's 21 Club brought Switzer's quest to a successful conclusion much sooner.

A letter written by Reeves at the time of Switzer's retirement from the Smithsonian offered his perspective on that afternoon of drinks and conversation on October 4, 1965: "I am a salesman, but he is a better salesman. In less than two hours at New York's 21 Club, he talked me out of a stone that I had intended my granddaughter and I die owning." What did the trick, aside from Switzer's passion and curatorial cunning? According to Switzer, Reeves was swayed by the idea of the gem being seen by generations of visitors and forever being known as the Rosser Reeves Ruby. The adman simply could not pass up such a wonderfully alliterative legacy.

During the couple of months before Reeves sent the ruby to the Smithsonian, he clearly was having fun with his new gem. When the initial October shipping date was delayed a month, he told Switzer he was "delighted to have thirty more days to caress this beautiful jewel." The ruby was finally conveyed to the Smithsonian in early December 1965, with a note from Reeves lamenting that: "As delighted as I am to see such a splendid museum get this remarkable stone, I must say that it is with a feeling of regret. It is my 'baby' and I miss it." Switzer, in his accession

OPPOSITE

Ruby is the gem name for the mineral corundum (aluminum oxide) that has been tinted red by tiny amounts of natural impurities of the element chromium. Light reflects off of three sets of parallel tiny, needlelike crystals of the mineral rutile, which naturally formed inside of the ruby, intersecting to form the six-rayed star. The threefold symmetry of the corundum atomic structure constrains the rutile crystals to align with equal probability in three directions at 120 degrees to one another.

memo to Smithsonian secretary S. Dillon Ripley, indicates regarding the ruby: "To the best of my knowledge, it is the largest and finest in the world."

As with many iconic gems, the story of the Rosser Reeves Star Ruby has accumulated a certain amount of romantic lore, which is fun and fascinating and good for marketing, but mostly has no basis in fact. There is the (tall) tale Reeves told to the *Arkansas Gazette*, of visiting the Middle East and being awakened to the news that "a great ruby has been found," after which he quickly flew to Istanbul, arriving just in time to purchase the gem at auction. It was perhaps irresistible for a salesman to not fill in some of the stone's unknown early history. Additionally, there are unverified accounts, supposedly by Reeves, of how he accidentally left "his baby" on the bar of the 21 Club before leaving on an overseas flight, and when finally able to call the Club upon landing, was relieved to learn that it was safely stored, awaiting his return. Another story describes how Reeves, who always carried the ruby in his pocket as his lucky charm, lost the stone while riding in a limousine, and it was found miraculously perched on the vehicle's undercarriage, where it had come to rest after slipping through the seat cushions.

Mackay Emerald Necklace

When telegraph and cable executive Clarence Mackay married Anna Case, former prima donna of the Metropolitan Opera, in July 1931, this art deco Cartier necklace, featuring a stunningly huge emerald, was his wedding gift to his bride. The velvety green central emerald contains a garden of veils and feathers and other inclusions, indicating it came from the Muzo mine in Colombia. The elegant and commanding necklace is set with more than two thousand diamonds and thirty-five emeralds, including the 167.97-carat faceted emerald in a pendant with baguette diamonds that measures 5×5.8 centimeters (2×2.3 inches). An intricate chain of alternating links of pavé-set and baguette diamonds is highlighted with two oval insets featuring emeralds weighing 4.77 and 4.35 carats, respectively. The heavy pendant has a pin backing so that it can be secured to the gown on which it is being worn. As was the case for many of the important emeralds in Cartier jewelry of the period, the large stone was likely mined when the Spanish controlled the Colombian emerald mines in the sixteenth and seventeenth centuries. Many of the emeralds went to India, where they were popular with the Moguls, who fashioned them into spectacular jewels. During the early twentieth century, Cartier conducted a lucrative business resetting Mogul-cut emeralds and other gems into contemporary designs for many Indian maharajas.

OPPOSITE
The Mackay Emerald Necklace, set with over two thousand diamonds and thirty-five emeralds, including the 167.97-carat faceted central emerald, is an outstanding example of the great 1920s art deco designs by Cartier.

Anna Case was born in 1889 in Clinton, New Jersey. The daughter of a blacksmith, she taught herself violin and organ and sang in her church choir. A local voice teacher encouraged her natural singing talent, and when Ms. Case debuted as a lyric soprano with the New York Metropolitan Opera in 1909, at age twenty, she was the only American singer without European training who had been accepted by the Met. She retired from the Metropolitan Opera in 1920 and continued her music career by giving concert recitals in the United States and abroad. She also performed public tone tests with Thomas Edison. He would challenge audiences to distinguish between his recording and the live singer, as a way of promoting his new phonograph.

While attending a performance at the Metropolitan Opera in 1916, wealthy telegraph tycoon and financier Clarence Mackay was enchanted by the young and beautiful Anna Case, and he later invited her to perform for a party at his opulent Long Island mansion, Harbor Hill. This was the beginning of a fifteen-year friendship and courtship, during which he wooed her with jewelry and carriages of flowers, that finally resulted in their 1931 wedding. (Mackay was divorced from his first wife, Katherine Duer Mackay, but his Catholic faith prevented him from remarrying while she was alive, hence the long courtship.)

Clarence Mackay's family came into their fortune when his father discovered the Comstock Lode near Virginia City, Nevada—the largest single deposit of gold and silver ever found, producing ore worth over $2.5 billion in today's dollars. Clarence Mackay became the chairman of the board of the Postal Telegraph and Cable Corporation and of the Mackay Radio & Telegraph Company. He oversaw the laying of the first communication cables across the Atlantic and Pacific Oceans.

Mackay's first wife died in 1930, and he and Anna Case were married a year later. By then, however, his financial situation had been greatly diminished by the 1929 stock market crash. Despite his money woes, Mackay was determined to give Anna an extravagant wedding gift, and he turned to his good friend Pierre Cartier. Many of Cartier's wealthy clients were affected by the stock market crash and depression, and consequently, numerous jewelry orders were canceled and new sales were exceedingly slow. Because of their friendship, and Cartier's confidence that Mackay's fortune would improve, and therefore a future sale was better than no sale, Cartier assisted Mackay in selecting the emerald necklace and gave it to him on credit. Cartier's faith in Mackay was well placed, as the $100,000 price was paid off in four installments by the end of 1932.

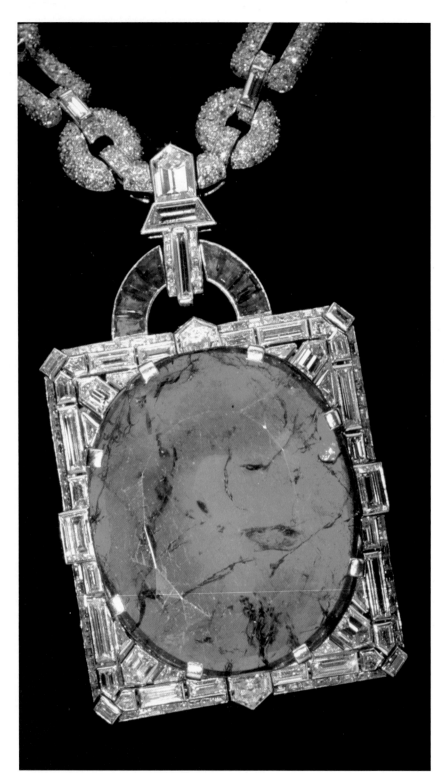

LEFT
Detail of emerald pendant.

ABOVE
Close-up image of the large emerald
reveals an inclusion of a crystal of
the mineral parisite, which confirms
the origin of the emerald as the Muzo
mine in Colombia.

Anna Case and Clarence Mackay outside of the Metropolitan Opera in 1931; she is wearing the emerald necklace.

Anna Case Mackay curtailed her public singing career when she was married, but she retained her celebrity status into the 1940s. She sang at charitable events and performed for servicemen in New York during World War II. She also performed on radio, notably in a 1938 Golden Rule Mother's Day special, which also featured the silent film star Mary Pickford (who bequeathed the Star of Bombay sapphire to the Smithsonian National Gem Collection).

After only seven years of marriage, Clarence Mackay died in 1938, and Anna never remarried. She lived in her New York apartment and remained a patron of the arts until her death in 1984. In her will, she bequeathed her spectacular emerald and diamond necklace to the Smithsonian Institution. It arrived at the museum in a white leather fitted case, inscribed ACM, July 18, 1931 (her wedding day). Curiously, a note that came with the necklace states that according to family lore, the necklace originally was created for the Russian royal family, but there is no documentation to support this claim.

Star of Asia

Renowned for its impressive size, intense color, and sharp star, the Star of Asia, which weighs 329.7 carats, is one of the world's finest star sapphires. Sapphires and rubies are gem varieties of the mineral corundum. The star forms when titanium atoms are trapped within the growing corundum crystal. As the crystal cools, the titanium forms needlelike crystals of the mineral rutile, which orient themselves in three directions. When properly cut, light reflecting off the three sets of needles produces the six-rayed star. This phenomenon is called asterism.

ABOVE
A close-up view of the center of the star reveals needlelike crystals of rutile oriented in three directions. Three bands of light reflecting off the rutile intersect to form the star.

RIGHT
Syed Mohamed Maricar and his partners, Edward and William Hopkins, admiring the large sapphire at their consignment house in London, circa 1958.

OPPOSITE
The 329.7-carat Star of Asia sapphire, with its impressive size, well-formed star, and translucent deep-blue body color, is one of the finest known.

The Smithsonian acquired the stunning Star of Asia sapphire from well-known mineral dealer Martin Ehrmann, who was representing owner Jack Mason of Los Angeles and Geneva. Received in exchange for a parcel of small faceted diamonds (which had been transferred to the museum from the General Services Administration), the great gem is considered to be the finest and most beautiful large star sapphire in the world. It came with a purported provenance that any collector would love, complete with an exotic source—Burma—and royal pedigree of having belonged to the maharaja of Jodhpur. Although the stone truly is magnificent, its history turns out to be an equally magnificent fabrication. Imagine, embellishing a story to help sell a gemstone!

So, what is the real background of the Star of Asia sapphire? A few years ago, we discovered a series of newspaper articles from August of 1950 announcing that the world's largest star sapphire, weighing 329 carats, was dazzling visitors at the first U.S. International Trade Fair in Chicago. Photos showed a model holding a silver dollar next to the gem to show off its astonishing size. According to the article, the stone was co-owned by Syed Mohamed Maricar of the Kohinoor Trading Company and King's Jewelers of Colombo, Sri Lanka. We contacted the grandson of Mr. Maricar, who confirmed that the sapphire had been mined in Sri Lanka and that his grandfather's company had owned the stone. He provided a copy of a June 1952 letter in which Fred Pough, curator of gems and minerals at the American Museum of Natural History, describes the 329-carat stone "to be unparalleled among star sapphires in its rich color, star perfection, and freedom from flaws." He also shared a photo, circa 1958, of his grandfather and his English partners, Edward and William Hopkins, admiring the large sapphire at their consignment house in London. Apparently, sometime later the stone made its way to Mr. Mason and Mr. Ehrmann, and then to the Smithsonian—with its newly embellished history. The truth is, this star sapphire is magical enough without need of historic trappings, but it is nice that we now know that it was found sometime before 1950 in Sri Lanka and did have a "star" turn in Chicago before becoming a centerpiece of our National Gem Collection.

Star of Bombay

The 181.8-carat Star of Bombay sapphire is from Sri Lanka. The stone measures 36.30×28.82×17.55 millimeters (1.4×1.1×0.7 inches), which is about the size of walnut. It is notable for its large size, exceptional transparency, rich blue color, and well-formed star. It was a bequest to the National Gem Collection in 1980 from silent film star Mary Pickford.

ABOVE
Mary Pickford wearing the Star of India, now Star of Bombay, necklace and the original Star of Bombay sapphire ring. An article, "Hollywood's Own Jewels," in *Vogue* in 1938 described Mary Pickford's 200-carat Star of India and 60-carat gems as "two of the purest sapphires in the world, but rarely are they unlocked from their vault."

OPPOSITE
The 181.8-carat Star of Bombay sapphire was mined in Sri Lanka prior to 1935.

Mary Pickford had her first supporting role on Broadway at age fourteen in 1906, and during the next ten years she appeared in hundreds of silent films and became known as "America's Sweetheart." She was one of the most recognized women in the world. She married actor Douglas Fairbanks Sr., and with Charlie Chaplin and D. W. Griffith, they founded United Artists Corporation, which allowed her artistic and financial control over her film projects. She made her last silent film in 1927, and after appearing in several sound films, she retired from the screen in 1934. Her marriage to Douglas Fairbanks ended in 1936, and the following year she married Buddy Rogers. At some point between those two events she visited her favorite jeweler, Trabert & Hoeffer, and a contemporary newspaper account (*Detroit Free Press*, June 14, 1936) mentions her purchase of two magnificent star sapphires—the Star of Bombay and the Star of India—for more than $100,000. How better to move on from a failed marriage than to buy yourself some big sapphires! Many published accounts report that the Star of Bombay was a romantic gift from Douglas Fairbanks, but it simply is not true. With Mary Pickford's business acumen and financial success, she could afford to buy her own jewelry. Buddy Rogers confirmed in a letter to a former Smithsonian curator that, regarding the sapphire, "she loved playing with it" and he was sure she owned it before he met her.

Trabert & Hoeffer was one of the hottest jewelers in Tinseltown in the 1930s, in part because it was one of the first New York jewelers to open a boutique in Los Angeles. Especially popular were their art deco creations set with huge cabochon gems, which showed up on screen and out on the town. In an inspired marketing move, they named many of their large gems as if they were part of a royal collection. A 1936 Trabert & Hoeffer advertisement announced, "The famous Star of Kashmir heads the finest collection of star sapphires," and another in 1935 proclaims of a 60-carat sapphire in a ring: "In all the world—the only one Star of Bombay."

There seems to be no doubt that the sapphire purchased by Mary Pickford as the Star of Bombay was a 60-carat stone set in a ring, and not the 182-carat gem in the Smithsonian collection. Rather, her original "200-carat" Star of India is, in fact, the stone we know today as the Star of Bombay. Why the switch? The transfer of names might have resulted from the realization by Mary Pickford that there was another Star of India sapphire—the giant 563-carat stone that had been on display at the American Museum of Natural History for more than thirty years. Perhaps, rather than

wearing another Star of India, she started referring to her larger and more noticeable sapphire as the Star of Bombay. When the 182-carat star sapphire came to the Smithsonian in 1980, as a bequest after Mary Pickford's death in 1979, the appraisal provided by the estate listed the gem as the Star of Bombay.

In a *Los Angeles Times* article (February 25, 1996), Edward Stotsenberg of the Mary Pickford Foundation recalls when the Smithsonian curator came to California to pick up the Star of Bombay: "Upon seeing the gem said: 'We want this sapphire. It is much brighter than our others.' He opened his pocketknife and pried the clasp loose, freeing the sapphire from the elaborate diamond necklace. He wrapped it in a soft cloth and put it in his coat pocket and returned to Washington." In the curator's defense, at that time, the collection focus was on gems rather than jewelry. Since then, our acquisition philosophy has evolved, and we appreciate the jewelry as an integral part of the story of the gem.

Victoria-Transvaal Diamond

In 1950 or early 1951, an extraordinary 240-carat diamond was unearthed in the Premier mine in South Africa's Transvaal region. Some accounts indicate that the miner who found it received a bonus equivalent to about a hundred dollars. The diamond crystal was acquired by Baumgold Brothers Inc., a major diamond importing and cutting firm in New York, and by mid-1951 was cut into two spectacular pear-shape champagne-colored gems. The smaller 64.07-carat stone, dubbed the Natasha Diamond, was sold almost immediately to Middle Eastern royalty, and then later to a princely family in India, before landing with a California gem dealer in 1988. The larger 75-carat sibling stone was a blaze of fire with 114 facets (most modern brilliant-cut diamonds have only 58 facets) and was christened the Transvaal Diamond after its place of discovery. Baumgold Brothers recut the diamond to improve the proportions and enhance its brilliance, reducing the weight to 67.9 carats. The fancy yellow-brown-colored diamond measures 32.51×24.42×14.36 millimeters (1.3×1×0.6 inches)—about the size of a walnut. The color is due to trace amounts of nitrogen substituting for some of the diamond's carbon atoms.

OPPOSITE

The champagne-colored 67.9-carat Victoria-Transvaal Diamond is a pear-shape brilliant cut with 116 facets and is suspended from a gold and platinum necklace comprised of sixty-six round brilliant-cut diamonds, fringed with ten drop motifs, each set with two marquise-cut diamonds, a pear-shape diamond, and a small round brilliant-cut diamond. The 108 diamonds in the necklace total 44.67 carats.

Pre-internet and cable TV, the options to promote a new diamond were: Take it on the road, attract attention of the newspapers, and make appearances on TV and in the movies. Well, Baumgold Brothers did all three. By October 1951, the "famous Transvaal Diamond" was the star attraction at the opening of a new diamond department at Horne's Department Store in Pittsburgh. It was mounted in a simple platinum necklace owned by the Guild of American Diamond Cutters and worn to a benefit in February 1952 by actress Arlene Dahl.

In October 1955, the diamond was displayed (unset) at two Sears stores in the Miami area. By March 1956, it had become the dazzling centerpiece of the current necklace. The stunning Transvaal Diamond necklace attracted even more attention, appearing alongside the Kimberley Diamond (also now part of the Smithsonian National Gem Collection) that spring at the Sears "Million-Dollar Diamond Sale" in Los Angeles. The *Los Angeles Times* effused that the exhibition was "one of the largest shipments of diamonds ever to be flown into Los Angeles at one time." The Transvaal Diamond was featured at the Brussels World Fair in 1958. It was reunited with the Kimberley Diamond in 1966 in South Africa at the Diamond Centenary Show, and again for an appearance in an episode of the popular 1960s television show *Ironside*.

Like many great gems, there is a bit of intrigue in the Transvaal Diamond's past. After making an appearance at the Tucson Gem and Mineral Society Show in February 1968, a Tucson jeweler had the necklace on consignment from Baumgold Brothers. He entered into an agreement with John Battaglia, who was suspected of having connections to organized crime, to sell the diamond and other jewelry pieces in Southern California. Shockingly, Battaglia left town with the goods and refused to pay the jeweler or return the jewels. The overwrought jeweler took his own life. Sometime after Battaglia died of a heart attack in 1971, the diamond and other jewelry resurfaced and was returned to Baumgold Brothers. In the aftermath of this harrowing episode, Baumgold Brothers put the Transvaal Diamond up for auction in March 1976 at the Desert Auction Galleries in Palm Springs, California. Twelve prequalified buyers were invited to the closed auction, including Leonard and Victoria Wilkinson from Prineville, Oregon.

Leonard Evert "Wilky" Wilkinson graduated from high school in 1932 in Mount Carmel, Pennsylvania, and married his local sweetheart, Victoria Dempnock, just a few years later. They operated a corner grocery store for several years, but Leonard's

dream was to own a lumber mill. In 1958, he started the Coin Millwork Company in Prineville, Oregon, making door and window frames and millwork. The business quickly grew to the largest millwork operation in one location in the world.

Leonard sent Victoria to the Palm Springs auction of the Transvaal Diamond, even though she told him that she did not want the gem, saying: "It is too splashy" and "We're just members of the common herd." The auctioneer proclaimed the auction was the most exciting thing to hit Palm Springs since Spiro Agnew. The bidding started at $200,000, and when the price hit $400,000, only three bidders remained; Victoria won the diamond for $430,000. At the time, it was the third largest sale of a single piece of jewelry ever recorded in the United States. Newspaper accounts state that only a week later, Leonard claims to have received offers to buy the stone for $550,000, but he refused to sell.

The Wilkinsons had other plans for the Transvaal Diamond. During the next year, they completed arrangements to donate the diamond, along with several other important jewelry pieces, to the Smithsonian National Gem Collection. It is not certain what inspired the gift, but it was formalized following a visit to their home in Reno, Nevada, from Smithsonian curator Paul Desautels. The Transvaal Diamond had been on his radar since seeing it exhibited years earlier at the American Museum of Natural History, and the publicity following the diamond's auction almost certainly rekindled his enthusiasm for the stone. Mr. Wilkinson's failing health (he died the year after giving the diamond) was also likely a contributing factor. But perhaps he was moved by the opportunity for a grand gesture to honor his beloved wife. The only condition specified by Leonard Wilkinson was that the Smithsonian would "name it the Victoria-Transvaal, in honor of my wife." The gift was accepted in 1977 from Leonard E. Wilkinson and Victoria Wilkinson.

Portuguese Diamond

From 1951 through 1957, this spectacular octagonal-shape 127.01-carat diamond was a featured attraction in Harry Winston's Court of Jewels exhibition that traveled throughout the United States. Winston publicists described the diamond as having the unique soft, warm haze usually found in the finest Brazilian diamonds. Furthermore, they reasoned, since all the important colonial Brazilian diamonds had become the property of the conquering Portuguese Crown, it is likely that this diamond was a former crown jewel. Therefore, they christened the stone the Portuguese Diamond. It was with this name that the diamond was acquired by the Smithsonian in 1963—and is how it is still known to the world today. As it turns out, however, the diamond is not from Brazil, and there is no connection to Portugal. Perhaps more appropriately, we should be calling this great diamond the Black, Starr & Frost Diamond or the Peggy Hopkins Joyce Diamond.

The venerable jewelry house of Black, Starr & Frost was founded in 1810, and at the beginning of the twentieth century was known in New York as the "diamond palace of Broadway." Fittingly, when a large diamond was discovered at the Premier mine near Kimberley, South Africa, in about 1910, it was quickly acquired by Black, Starr & Frost. Initially formed into an approximately 150-carat oval cushion shape, by the time it was brought into New York in 1912, it had been recut into the familiar 127.01-carat octagonal-shape gem we know today. In the early 1900s, the clean, modern shape had recently become fashionable, especially for larger diamonds. Magazine advertisements in 1913 and 1914 by Black, Starr & Frost showed the diamond and described it as the finest known blue-white diamond. An advertisement in *Vanity Fair* (January 1924) showed a drawing of the unset diamond and proclaimed: "Another historic jewel is the Black, Starr & Frost diamond. It is a blue diamond, of a particular intensity of color, and weighs 127 carats—larger than the Koh-i-Noor. More than that, it is the largest blue diamond ever discovered. It is absolutely perfect in every way, and it is the largest diamond of any kind which is offered for sale." The price was $300,000. (Note: The blue color referred to in the description is the result of the diamond's intense blue fluorescence, which is visible in sunlight and under certain indoor lights; the body color of the diamond is actually very light yellow.). It is not surprising that such a diamond should catch the eye of the much-publicized collector of wealthy husbands and large jewels, Peggy Hopkins Joyce.

Marguerite "Peggy" Upton was the daughter of a barber and was born near Norfolk, Virginia, in 1893. By age seventeen she joined a vaudeville troupe, and by age twenty-four she had been married twice, expertly navigated the wilder side of both Washington, DC's and New York's social scenes, landed a starring role as a celebrated Ziegfeld Follies girl, and had small parts in films and on Broadway.

By all accounts, Peggy Hopkins was not a talented performer, but with her natural beauty and audacious lifestyle, she was a sensation. Peggy and the tabloids were made for each other: Her many marriages, divorces, and very public love affairs were daily fodder for the *New York Daily News*. In fact, her public persona was the inspiration for the heroine in Anita Loos's novel *Gentlemen Prefer*

LEFT
Peggy Hopkins Joyce with the Portuguese Diamond choker necklace.

OPPOSITE
The 127.01-carat Portuguese Diamond is the largest faceted diamond in the National Gem Collection.

LEFT
Portrait by Raymond Perry Rodgers Neilson of Peggy Hopkins Joyce with her 127.01-carat diamond necklace.

BELOW
The Portuguese Diamond was temporarily set into a hatpin by Harry Winston Inc. for this photograph of actress Michelle Pfeiffer for *Life* magazine in 1995.

When exposed to ultraviolet light, the Portuguese Diamond fluoresces bright blue. The fluorescence is so intense that it is visible even in bright sunlight, prompting some early descriptions to refer to it as a blue-colored diamond.

Blondes, and she was mentioned in Cole Porter songs, an Irving Berlin musical, and punch lines in Will Rogers's monologues. In her memoir, *Men, Marriage and Me*, she claimed to have been engaged to more than fifty men in her life!

When she met husband number three, lumberman James Stanley Joyce, the new Mrs. Joyce took full advantage of her husband's wealth, going on lavish shopping sprees, and at one point convincing him to spend $300,000 on an extravagant pearl necklace instead of a yacht. Peggy and Joyce divorced in 1921, and although she had to give up their Miami mansion and received no alimony, she was able to keep most of her jewelry.

By 1928, Peggy had added a Swedish count to her list of ex-husbands. In March of that year, she walked into the offices of Black, Starr & Frost and purchased the $300,000 emerald-cut diamond that we now know as the Portuguese. There was speculation at the time that the jewel had been purchased for her by Walter Chrysler, with whom she was having an affair, but statements from the jeweler indicate she acquired the diamond by trading a $350,000 string of pearls, presumably those from Mr. Joyce, and paying $23,000 in cash. The 127-carat octagonal-shape diamond was set in a stunning choker-style platinum necklace with three rows of 383 small diamonds. The purchase created a news sensation, with columnists referring to the massive stone as "Peggy's skating rink." The *New York Times* described it as the world's largest diamond owned by a single individual, and noted that each time she wore the necklace, a private detective, as required by the insurance company, was in attendance. She remarked to a journalist: "I sometimes think it is more worry than a husband."

With the onset of the Great Depression, the spotlight on Peggy Hopkins Joyce inevitably dimmed. She married twice more, but to little fanfare. During the late 1940s, her great diamond necklace made appearances at special expositions and jewelry stores around the country, and was advertised for sale as the world's largest emerald-cut diamond. The necklace attracted considerable press, and the stories indicated that the diamond was originally cut in Europe; in one case it was said to have been owned by an Indian potentate. Interestingly, no mention was made of Peggy Hopkins Joyce, who apparently was still the owner. Harry Winston purchased the gem from Peggy in October 1951, and it traveled as part of his Court of Jewels exhibition until 1957. In 1963, he traded the diamond to the Smithsonian for a parcel of small faceted diamonds.

As for Peggy, who died at age sixty-four in 1957, she was perhaps comforted by the thought that her great diamond's history would record that it was "first worn by Peggy Hopkins Joyce." A *New York Times* article, written after she purchased the diamond, noted that, unlike the famous Hope Diamond, the stone had no romantic past of its own—and went on to say that Peggy planned to make one for it. Which she did. Although the diamond is called by a different name, its story forever will ensure that Peggy Hopkins Joyce is remembered.

Warner Crystal Ball

The amazing Warner Crystal Ball is the largest known flawless quartz sphere—a true wonder of nature. It measures 326.7 to 326.9 millimeters (12.861 to 12.872 inches) in diameter and weighs 48.5 kilograms (106.7 pounds). It was cut and polished from a massive quartz crystal in 1923 in Shanghai. Mrs. Cornelia Warner gifted it to the Smithsonian Institution in March 1930: "presented to the United States National Museum in memory of Worcester Reed Warner by Mrs. Worcester Reed Warner." It has been exhibited at the Smithsonian National Museum of Natural History since 1925. And there it has remained, one of the Smithsonian's most popular icons, but if not for the enthusiasm and generosity of Mr. Warner and later his wife, the great crystal ball might have been lost.

OPPOSITE
The Warner Crystal Ball is the largest known flawless quartz sphere. It measures 326.7 to 326.9 millimeters (12.861 to 12.872 inches) in diameter and weighs 106.7 pounds.

In November 1924, the Smithsonian Institution received an intriguing letter from Worcester Reed Warner of Tarrytown, New York, in which he stated that there was "a possibility, I don't even call it a probability, that the largest perfect crystal ball may be offered as a present to the National Museum." Warner was a noted mechanical engineer and astronomer and retired president of Warner & Swasey Company of Cleveland, which had designed and built the great telescopes at the Lick and Yerkes observatories, as well as others in Canada and Argentina. He was also a connoisseur and collector of fine Asian art. In a follow-up letter in January 1925, he announced that "the crystal ball is now in the Custom House in New York, having arrived there last week. The owner Mr. Lee Van Ching is trying to get data from China proving it to be over one-hundred years old. It is 13¼ inches in diameter and said to be without flaw and perfect." Two days later he wrote: "The customs officials here, in view of the fact that it cannot be duplicated at any price, have appraised it at $150,000 and demand fifty percent tariff. This makes me smile, but the Orientals were sober. Mr. Lee expects to take it back to China." In early February, a determined Warner suggests to the Smithsonian secretary, Charles Walcott: "I believe an urgent word from you to the treasury department would lead them to advise making the rules 'flexible' to the extent of allowing the National Museum to receive this exhibit for the minimum time of three months. Lee Van Ching, the owner, and Otto Fukushima, the agent, have agreed to a temporary loan of the ball to the museum." Walcott's subsequent request for the loan from the secretary of the treasury was approved under the provision that the ball was permitted to enter for exhibition by the Smithsonian Institution. It was released to the Smithsonian and by May 1, 1925, had been placed on exhibition in the National Museum's mineral hall.

According to customs correspondence, the crystal ball was brought into New York on November 19, 1924, under an appraisement entry by Lee Van Ching, and attached to the consular invoice was a certificate of antiquity alleging the ball to have been produced more than one hundred years ago—thereby avoiding import duty. Customs initiated an inquiry into the antiquity claim and held the ball in lieu of payment of $75,000 import duty; it was during this discovery period that the crystal ball was loaned to the Smithsonian. The report from the customs attaché in Shanghai disclosed that the ball was "not antique but thoroughly modern, having been cut and polished from the rock crystal in 1923." The intrepid investigator provided an impressively detailed description

of the making of the crystal ball. The rough rock was brought to Shanghai during August 1923 from Rangoon, Burma, by Wang Zong Teng, who traveled between Shanghai, Hong Kong, and Rangoon in pursuit of general commission business. The rock mass was entered in the Maritime Customs and duty paid therein; the weight was 950 kin (~1,260 pounds). The rough crystal was purchased by Lee Van Ching for about $20,000 Mex. (~$10,400; China then used Mexican silver "dollars" equivalent to about 0.52 U.S. dollars), and it was turned over to a jade cutter, Zai Yu Yue, located in Native City, Shanghai, who cut it to a rough ball, measuring 9.1 Chinese inches (~13.3 inches) in about three months, at a cost of $800 Mex. (~$415). The rough ball was sent to Shing Hwe Optical Co. in French Town, Shanghai, where it was finished and polished at a cost of $500 Mex. (~$260). The finished ball was taken to Lee Van Ching's establishment in French Town, where it was kept under cover in his sleeping apartment.

Upon Ching's return to Shanghai from New York during late summer 1924, he stated he had a prospective customer in New York for the ball and prepared to ship it. He directed his secretary, Ting Pao Chen, to prepare a consular invoice and certify the same with a letter of antiquity. The ball was smuggled out of Shanghai in the baggage of Kong N. Chow, who was a passenger on the S.S. *President Jefferson*.

Meanwhile, the crystal ball had settled into its new surroundings at the National Museum. Warner shared his excitement with Secretary Walcott in May 1925: "I am delighted that it is landed in the National Museum, and I believe it will stay there. Nature made it and man polished it. Nature did a unique job and the National Museum has it on exhibition. Don't let it get away for a thousand years, cannot replace or duplicate it." In response to the request from Smithsonian secretary Charles Walcott, in August the treasury department indicated that the ball was considered to be on permanent exhibition and was charged to the Smithsonian and may remain on exhibit indefinitely.

The fact was, however, that the crystal ball was still the property of Lee Van Ching and was only temporarily under the charge of customs, pending payment of the assessed duty or other resolution. The Smithsonian made overtures to Ching's agent in attempts to purchase the ball outright, alluding to the denied antiquity claim and large duty, but it could not afford the asking price. In 1928, the owner requested return of the ball within two weeks to send it to England for exhibition and to possibly find a buyer. Customs officials determined that the ball could be exported by the owner, but fortunately the plan was not carried out. Realizing the precarious situation for the crystal ball to remain at the museum, Assistant Secretary Alexander Wetmore reached out to Warner: "I hope you are in a position to present the object to the National Collection. . . . We consider that the sphere is one of the most attractive and important objects that we have anywhere in our exhibits." In February 1929, Warner negotiated a deal to purchase the ball from Fukushima, acting for Ching.

Worcester Reed Warner died in June 1929 while traveling in Germany. Later that year his widow, Mrs. Cornelius Warner, wrote to inform the Smithsonian that her husband had kept a special fund amounting to the remainder of the purchase price for the ball, contingent upon obtaining a concession from customs for admitting the ball duty-free. After discussions with the Smithsonian secretary, customs officials negotiated a settlement whereby Mr. Fukushima would pay to the treasury department $10,000 of his received payment for the ball (no, they do not forget!). With this agreement executed, Mrs. Warner completed the purchase of the crystal ball and the secretary of the treasury released the ball to the National Museum.

OPPOSITE
The Warner Crystal Ball, shown here, circa 1930, has been exhibited at the Smithsonian National Museum of Natural History since 1925.

Warner Crystal Ball

Hooker Emerald

The 75.47-carat Hooker Emerald is extraordinary for its vibrant green color and its exceptional clarity for such a large (27.1×26.9 millimeters; 1.07×1.06 inches) stone. The emerald originally was mined in Colombia, and likely was shipped to Europe by the Spanish during the sixteenth or seventeenth century. Emeralds were especially popular with the Moguls in India and among the sultans of the Ottoman Empire. The emerald was acquired by Tiffany & Co. at an auction of Sultan Abdul Hamid II's jewels. Janet Annenberg Hooker purchased the emerald brooch in 1955 and donated it to the National Gem Collection in 1977.

ABOVE
The centerpiece of this 1939–1940 World's Fair House of Jewels display is the tiara created by Tiffany & Co., featuring the magnificent 75-carat square-cut emerald it had purchased at the 1911 sale of the Turkish sultan Abdul Hamid II's property. The tiara could be separated into clips.

OPPOSITE
The 75.47-carat Hooker Emerald Tiffany & Co. brooch is surrounded by 109 round brilliant diamonds (10 total carats) and twenty baguette diamonds (3 total carats).

In November 1911 in Paris at the Galerie Georges Petit, there was an extraordinary auction of jewels that had belonged to the former Sultan Abdul Hamid II. The *New York Times* (November 26, 1911) described the exhibition room: "a wonderful spectacle, ablaze with lights reflected by the wealth of diamonds and precious stones of all kinds. . . . The collection of emeralds is most gorgeous. Some are as large as walnuts, and a few of them are of perfect color." The event attracted connoisseurs and representatives of major jewelry houses from around the world. After the first day of the sale, the *Times* reported that "emeralds were the principle feature at today's sale. . . . The most precious item was a brooch made of a large emerald, which sold for $25,000." The buyer of the emerald brooch (lot 46 in the *Catalogues des Perles, Pierreries Bijoux et Objets d'Art Précieux, le tout ayant appartenu a S. M. le Sultan*) was Tiffany & Co.'s London office, and the magnificent gem is the Hooker Emerald that today is in the Smithsonian National Gem Collection.

Abdul Hamid II became sultan at age thirty-four in 1876. He was the last supreme leader of the Ottoman Empire, ruling over a fracturing state at a time of great internal turmoil and numerous wars. Discontent with his autocratic rule and threats of European interventions in the Balkans led to a military uprising of the Young Turks in 1908, and the sultan was deposed in 1909. He spent his last years in exile and died in 1918. Before relinquishing power, he sent his jewels to Paris, hoping to use them as a source of income for his family. Instead, the agent overseeing the auction directed the proceeds to the Young Turks to be used to support their navy.

Tiffany & Co. set the 75.47-carat emerald as the centerpiece of a tiara consisting of five detachable clips, which could be worn assembled or individually, with 908 emeralds and diamonds weighing a total of 208.9 carats. The tiara was featured in the House of Jewels at the 1939–1940 New York World's Fair. The large emerald was subsequently removed from the tiara (apparently no longer the height of fashion) and mounted into the current brooch setting, which was introduced to the public in the 1950 Tiffany Blue Book (Christmas catalog).

Among those in the World's Fair crowds who admired the dazzling emerald tiara was Janet Annenberg Hooker, and in January 1955, she purchased from Tiffany & Co. the brooch with that same large emerald. Janet Annenberg was born in 1904 in Chicago, one of eight children. In 1920, the family moved to New York's Long Island. Her father, Moses, founded the Triangle Publishing empire that eventually included *Daily Racing Form*, the *Philadelphia Inquirer*, *Seventeen* magazine, and *TV Guide*. Janet's

LEFT
Sultan Abdul Hamid II (1842–1918) was said to have worn the emerald in his belt buckle.

BELOW
Page from catalog of 1911 auction of Sultan Abdul Hamid II jewels, showing at upper right the emerald brooch purchased by Tiffany & Co.

The Tiffany & Co. 1950 Christmas catalog featured the Hooker Emerald newly set in a brooch.

marriage to publisher L. Stanley Kahn in 1924 ended in divorce, and in 1938 she married investment banker Joseph A. Neff, who died in 1969. In 1974, she married James Stewart Hooker, who was head of labor relations for the *Philadelphia Inquirer*, and he passed away in 1976. Janet's brother, Walter Annenberg, took over the family publishing business in 1942. After great success in expanding the business, he sold most of it to Australian Rupert Murdoch in 1988, much to the benefit of his family, including Janet. Janet Annenberg Hooker's philanthropy extended to numerous cultural and government institutions. She was a major contributor to the Metropolitan Opera, and she supported significant remodeling projects at the White House and State Department. In 1977, she gave her Tiffany emerald brooch to the growing Smithsonian National Gem Collection, and in following years she encouraged her sisters Lita Annenberg Hazen and Evelyn Annenberg Jaffe Hall to also make major gifts to the collection. Janet provided the lead financial gift for the remodeling of the Hall of Geology, Gems, and Minerals at the Smithsonian National Museum of Natural History, which opened in 1997 and is named in her honor. And in 1994, she donated a spectacular suite of yellow diamonds to the collection in honor of her two sons, Gilbert S. Kahn and Donald P. Kahn.

Hooker Yellow Diamonds

The dazzling starburst-cut yellow diamond suite by Cartier consists of an 18-karat gold necklace set with fifty natural, fancy yellow diamonds, ranging from 1 to 22 carats, totaling 244.1 carats; clip earrings *each* featuring a 25.3-carat yellow diamond, surrounded by four pear-shape (3.33 total carats) and sixteen baguette-cut white diamonds (10.07 total carats); and a ring with a 61.12-carat modified brilliant-cut fancy yellow diamond accented by two 4.75-carat white trilliant-cut diamonds—in total almost 350 carats of fiery yellow gems. They were a gift to the National Gem Collection in 1994 from Janet Annenberg Hooker.

Although it seems Hooker never wore the jewelry suite publicly—due to her declining health—her granddaughter reported that Mrs. Hooker enjoyed wearing it when having tea in her apartment with her grandchildren.

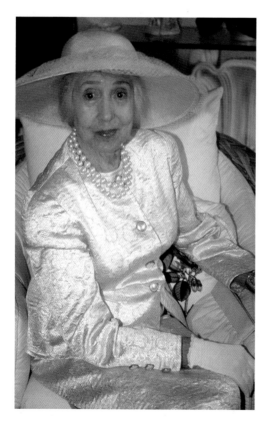

In the spring of 1994, I received a phone call from Janet Annenberg Hooker's son Gilbert Kahn. He invited me to join him at a Miami bank to view his mother's jewelry collection, offering that perhaps there might be something that would make an appropriate addition to the National Gem Collection. Even at the Smithsonian, such opportunities are not commonplace, and the next day, with enthusiastic anticipation, my colleague and I boarded a plane to Miami.

Gilbert met us at the airport and drove us directly to the bank. We were escorted to a secure room with a large table, upon which were placed several large safety deposit boxes. Once the bank officials departed and the door was locked, Gilbert removed the contents of the boxes, covering the entire table with all manner of jewelry cases—Cartier, Tiffany, Van Cleef & Arpels, etc. One by one he opened the boxes, revealing his mother's lifetime of jewels—a dazzling array of brooches, necklaces, and rings. When finished, he announced that his mother was wondering if any of her pieces might be worthy additions to the Smithsonian National Gem Collection. He also indicated that his mother planned eventually to give the remaining pieces to family members. Although still mesmerized by the spectacle of a table of sparkling gems, we began the task of focusing on the individual pieces. There were many spectacular designer pieces, but I was mostly interested in the jewelry with significant gems. Immediately, my attention was drawn to a bright yellow glow from one corner of the table. There, standing out from all the rest, was a Cartier suite featuring the most amazing yellow diamonds I had ever seen, and they were huge! With their starburst cut, they were ablaze beneath the vault lights. Gathering my composure, I quietly asked Gilbert if these yellow diamonds might be an option for the Smithsonian collection. He explained that his mother only recently (1989–1990) had purchased these pieces. She had always admired yellow diamonds, and she first noticed the ring at Cartier near her home in New York and bought it. She later added the earrings and necklace. He thought that his mother would be delighted to give the yellow diamonds to the Smithsonian, but he would confirm with her and let us know. We returned Mrs. Hooker's jewels to the safety deposit boxes and he dropped us back at the airport. For this Smithsonian curator, it was very good day!

Several days later, Gilbert confirmed his mother's intention to donate the starburst yellow diamonds to the National Gem Collection in honor of her sons (Gilbert and his brother, Donald P. Kahn), where they would join the spectacular 75-carat emerald she had given in 1977 and be available for the opening of the Janet Annenberg Hooker Hall of Geology, Gems, and Minerals in 1997.

Hall Sapphire Necklace

The spectacular Hall sapphire and diamond necklace highlights thirty-six cushion-cut Sri Lankan sapphires totaling 195 carats, accented by 361 round brilliant-cut diamonds (61.5 total carats) and seventy-four pear-shape diamonds (22.25 total carats). The many diamonds reflect light into the sky-blue sapphires and give them a luminous and airy look. The sapphires in the Hall necklace come out of a centuries-old tradition of mining in Sri Lanka: The country has been an important source of sapphires, rubies, and other gemstones for more than two thousand years. In Sri Lanka's central mountains, gems are still picked by hand from eroded alluvial gravel deposits that cover most of the southern half of the island. Sapphires from Sri Lanka are typically light to medium blue in color and can be very large in size. The necklace was created by Harry Winston Inc. for Evelyn Annenberg Jaffe Hall. Hall gifted the necklace to the Smithsonian in 1978. Hall's sisters, Janet Annenberg Hooker and Lita Annenberg Hazen, also donated jewelry pieces to the Smithsonian collection.

Hazen Diamond Necklace

The elaborate and intricate design of the Hazen diamond and platinum necklace is a classic example of the exquisite craftsmanship by Harry Winston Inc. from the 1960s. The necklace was gifted to the National Gem Collection in 1978 by Lita Annenberg Hazen, one of eight Annenberg siblings that included publisher Walter Annenberg, as well as sisters Janet Annenberg Hooker and Evelyn Annenberg Jaffe Hall, both of whom also donated major jewelry pieces to the Smithsonian National Gem Collection. Apparently, after giving her large emerald in 1977, Mrs. Hooker challenged her sisters to follow her lead. Mrs. Hazen was noted for her philanthropy for medical research and was a generous patron of the sciences and arts.

The Hazen necklace consists of 325 high-quality, well-matched diamonds, weighing 131.43 total carats, set in platinum that highlights multiple diamond shapes. The upper section is a single row of emerald-cut diamonds, and the lower section consists of a row of baguette-cut diamonds and a row of round brilliant-cut diamonds from which a fringe of pear-shape diamonds (the largest is 5 carats) are suspended. Typical of the versatility designed into many Winston pieces, the necklace can be separated into two sections and can be worn as either a classic single row of diamonds or as the four-row necklace.

75

Maharaja of Indore Necklace

This exquisite emerald and diamond necklace features some of the oldest faceted gems in the National Gem Collection. The fifteen emeralds and sixteen large football-shape diamonds were likely cut in the sixteenth or seventeenth century during the Mogul period in India. As with most emeralds fashioned by the Moguls, these originated from the mines in Colombia when they were under Spanish control from the late 1500s through 1675. The emeralds in the necklace are of the highest quality, and the 44.84-carat central emerald, with its rich velvety-green color and exceptional clarity, is among the world's finest. Typical of the Indian style, the emeralds are center-drilled to be attached by thread or wires to the necklace.

The emerald and diamond necklace was purchased in its present form by Harry Winston from the maharaja of Indore in 1947. Today, Indore is part of the Madhya Pradesh state of India. Maharaja Yashwant Rao Holkar II, son of Tukoji Rao Holkar III, ascended the throne of Indore in 1926 and ruled until Indian independence in 1947. It was not uncommon that former ruling families were forced to sell the family jewels to compensate for income lost when India became a republic, and jewelers such as Winston, Cartier, and others were ready buyers. Harry Winston featured the Maharaja of Indore Necklace in his fabulous Court of Jewels exhibition, which also included the Hope Diamond and several other renowned gems; it traveled around the United States from 1948 through 1953 to benefit the National Foundation for Infantile Paralysis (now known as the March of Dimes). Curiously, in the booklet describing the exhibition, Winston called it the Inquisition Necklace, with no explanation other than "it dates back to that period in history." He also suggests that it was first owned by Spanish royalty and later adorned ladies of the French Court before being acquired in the early 1900s by the maharaja of Indore. This European connection has not been documented and seems unlikely, considering the Mogul-style cuts for the large emeralds and diamonds. This would not be the only time that Harry Winston added creative embellishments to make a good story even better.

In 1955, Harry Winston sold the emerald and diamond necklace to Cora Hubbard Williams of Pittsburgh. Mrs. Williams was the daughter of John Winslow Hubbard, president of Hubbard and Company, which made axes, shovels, saws, and hoes, and of the Pittsburgh Ice Company, Hubbard Steel Company, and Mississippi Navigation Company. She married John Carmichael Williams and dedicated much of her time to helping children with developmental and educational needs. Upon her death in 1971, she bequeathed the necklace to the Smithsonian National Gem Collection. This arrangement likely was encouraged by Harry Winston, as in her will Mrs. Williams mentions that the necklace be added to the collection containing other gems given by Harry Winston.

LEFT
Maharaja Yashwant Rao Holkar Bahadur of Indore, 1930.

OPPOSITE
The necklace contains 360 diamonds and fifteen emeralds; the lower pendant and upper chains, set with platinum, were added to the antique strands in the twentieth century. The original double strand is decorated with eight pairs of large diamonds and four pairs of barrel-shape emeralds. The pendant is made up of five emeralds and twenty-four diamonds and is placed centrally on the double strand. The finial on the clasp echoes the design of the pendant and is also a modern addition to the necklace.

Ancient Cutting Style

The large emerald in the Maharaja of Indore Necklace is an example of a traditional Indian cutting style that goes back at least to the time of the Roman naturalist Pliny the Elder (first century CE). In *The Natural History,* Pliny describes the cutting of beryl gems (emerald is a green gem variety of the mineral beryl) in India: "They prefer, too, cutting the beryls in cylindrical form, instead of setting these as precious stones; an elongated shape being the one that is most highly esteemed. . . . The people of India are marvelously fond of beryls of an elongated form and say that these are the only precious stones they prefer wearing without addition of gold; hence it is that, after piercing them, they string them upon the bristles of the elephant."

ABOVE
The large diamonds and emeralds in the Maharaja of Indore Necklace were
fashioned in India in the seventeenth century. The large central emerald weighs
approximately 45 carats and is strung onto the necklace through a hole drilled
lengthwise down its center. The inside of the hole was polished to make it
less visible. The large emerald's elongated, prismatic form indicates that the
original crystal's faces were simply rounded off and polished to yield the largest
possible gem.

LEFT
Circa 1950, actress Katharine Hepburn modeled the Maharaja of Indore Necklace wearing her Rosalind costume from *As You Like It*.

BELOW
The sixteen large old-mine-cut diamonds suspended on the necklace are each drilled with two angular holes that connect in the stone, allowing a wire to pass through the stone and secure it almost invisibly to the necklace. These diamonds are slightly yellow and almost certainly were mined in India, which was the only commercial source of diamonds until deposits were discovered in Brazil in the 1720s.

Maharaja of Indore Necklace

Chalk Emerald

The vibrant green color and exceptional clarity and brilliance of the Chalk Emerald and its characteristic mineral inclusions indicate that it is from the famed Muzo mining region in Colombia. The original crystal likely was shipped by the Spanish during the sixteenth and seventeenth centuries to India, where these brilliant green stones were highly prized by the Mogul rulers. Oscar Roy Chalk purchased the emerald, in the current ring setting, from Harry Winston in 1962 for his wife, Claire. Winston had acquired the loose stone in 1959 through a broker who purchased it in India from the maharaja of Cooch Behar. Winston recut the emerald to improve its appearance, reducing the weight from 38.38 to 37.82 carats, before setting it into the gold and platinum ring. Mr. and Mrs. O. Roy Chalk gifted the emerald ring to the Smithsonian National Gem Collection in 1972.

Oscar Roy Chalk was the son of immigrants and raised in New York City. He attended law school, during which he met and married his wife, Claire; they were successful entrepreneurs in real estate, transit companies, and airlines. They started the *Washington Examiner* newspaper in Washington, DC, and two Spanish-language papers in New York. The Chalks had residences in New York, Washington, DC, and Palm Beach, Florida, and they owned an important art collection. He liked cars, yachts, and clothes; she was active in Palm Beach society. Mr. Chalk chaired United Nations finance committees and founded the American-Korean Foundation.

Before Harry Winston purchased the emerald, it was the centerpiece of an emerald and diamond necklace owned for centuries by the royalty of Baroda (present-day Gujarat in India). The daughter of Maharaja Sayajirao Gaekwad III famously wore the necklace to all royal functions. In 1911, she married Jitendra Narayan and became Maharani Indira Devi of Cooch Behar, a state in west Bengal. Her son, Maharaja Jagaddipendra Narayan Bhup Bahadur, reigned from 1932 to 1949 and was the titular maharaja until his death in 1970. After the formation of the Union of India, when the royal families lost their traditional status and income, the maharaja sold the emerald that eventually was purchased by Winston.

LEFT
Maharaja Jagaddipendra Narayan at his coronation, circa 1936, reigned as head of the Indian Cooch Behar state until 1949.

OPPOSITE
The 37.82-carat Chalk Emerald in a ring setting by Harry Winston Inc., with sixty pear-shape diamonds, totaling 15.62 carats.

Whitney Alexandrite

Both images are of the same alexandrite gem, but under different light. Alexandrite is a gem variety of chrysoberyl renowned for its dramatic and exotic color change from purplish-red under incandescent light to bluish-green in daylight or fluorescent light. Trace impurities of chromium in chrysoberyl cause alexandrite to transmit red and green light equally well. Tungsten lights or candlelight emit most strongly in the red portion of the spectrum, causing the gems to appear red. Sunlight and fluorescent light, on the other hand, are rich in green light, causing the alexandrite to appear green. The way that the human eye perceives color accentuates the effect. The 17.08-carat, cushion-cut Whitney Alexandrite from the Hematita mine in Minas Gerais, Brazil, is exceptional for its size, near-perfect clarity, and dramatic color change from raspberry red under incandescent light to teal green when illuminated by daylight.

The original locality for alexandrite is Russia; however, fine gems have also been found in Brazil, Sri Lanka, Tanzania, Madagascar, Zimbabwe (formerly Rhodesia), India, and Burma. Alexandrites over 2 carats are considered large, and when over 5 carats, they are extremely rare.

Nils Gustaf Nordenskiöld (1792–1865), a renowned Finnish mineralogist, is credited with the discovery of alexandrite. While examining a newly found mineral sample he had received from Count Lev Alekseevich Perovskii, which Nordenskiöld first identified as emerald. He was confused by the high hardness and then surprised to see that under candlelight, the color of the stone had changed to red instead of green. He confirmed the discovery as a new variety of chrysoberyl and suggested the name "diaphanite." Perovskii, however, used the rare specimen to ingratiate himself with the Russian imperial family. He renamed it "alexandrite" and presented it to the future tsar (Alexander II) in honor of his coming of age on April 17, 1834. However, it wasn't until 1842 that the description of the color-changing chrysoberyl was published for the first time under the name "alexandrite." (Information provided by Gustaf Arrhenius, 2010.)

The world-class alexandrite gem was generously gifted to the National Gem Collection in 2009 by Coralyn Wright Whitney. Dr. Whitney was a research professor in biostatistics, and after retiring from academia, she earned her graduate gemologist and accredited jewelry degrees from the Gemological Institute of America. Because of her knowledge and experience, she recognized that an alexandrite gem this rare and beautiful was truly special. Dr. Whitney generously supported several other Smithsonian educational and research programs, and she also donated the Whitney Flame Topaz to the National Gem Collection.

Whitney Flame Topaz

The Whitney Flame Topaz is named in honor of the donor, Dr. Coralyn Wright Whitney. Smithsonian curators and Dr. Whitney first encountered the stone at the 2018 Tucson Gem and Mineral Show, and when they saw the smoldering stone, they knew instantly it belonged in the National Gem Collection.

The topaz was recovered in the early 1970s from the Capão mine in the famous Ouro Preto area in Minas Gerais, Brazil. The rough crystal weighed more than 200 carats and was originally cut to just over 50 carats for a private collector. The gem was recut to the current 48.86-carat pear shape to improve its proportions. The large size, fiery red color (caused by trace impurities of chromium), outstanding clarity, and stunning brilliance make this one of the finest imperial topaz gems in the world.

Only 1 to 2 percent of the topaz mined in the Ouro Preto region is gem-quality. Imperial topaz is coveted for its golden-orange to orange-red hues, and pure red gems such as the Whitney Flame are exceedingly rare and highly prized.

Cullinan Blue Diamond Necklace

The Cullinan Blue Diamond Necklace is a beautiful example of the graceful and elegant Edwardian–style jewelry popular in the early twentieth century. The necklace is made of 9-karat rose-gold with silver top and set with 243 round, colorless diamonds and nine blue diamonds, totaling over 30 carats. It has a detachable double-ribbon bow motif brooch with a dangling pendant that features the 2.6-carat fancy intense-blue Cullinan Blue Diamond. The remaining eight blue diamonds total approximately 3 carats.

ABOVE
Thomas Cullinan (left) and Mine Manager Thomas McHardy with the Cullinan Diamond in 1905.

OPPOSITE
The necklace is set with 243 round colorless diamonds and nine blue diamonds, totaling over 30 carats. The pendant features the 2.6-carat fancy intense-blue Cullinan Blue Diamond.

According to family lore, when Thomas Cullinan started the Premier diamond mine in South Africa in 1903, he told his wife, Annie, that one day he would bring her the "biggest diamond in the world." In January 1905, his mine did, in fact, produce the largest diamond ever found, the fist-size 3,106.75-carat Cullinan Diamond. Instead of going to his wife, however, the stone was purchased in 1907 by the Transvaal government and presented to England's King Edward VII as a "token of the loyalty and attachment of the people of Transvaal to His Majesty's throne and person." The diamond was cut into nine large gems, which belong to the British Crown, and numerous smaller stones.

In 1910, Cullinan was knighted as part of the celebration of the inauguration of the Union of South Africa for his contribution to South African industrial development. At about this same time, he presented the Edwardian-style diamond necklace to his wife, perhaps as consolation for having lost out on the big diamond given to the king. The nine blue diamonds in the necklace, including the 2.60-carat Cullinan Blue Diamond, represent the nine major diamonds cut from the Cullinan Diamond. They are all from Cullinan's mine, which today remains the most important source of blue diamonds, and likely were among the first blue diamonds recovered from the mine.

The Cullinans gave the necklace to their eldest daughter, Winifred Solomon, and it later went to *her* eldest daughter, Wynne Herrick. Wynne was the last family member to wear it publicly: In 1967, she donned the necklace to accompany Lady Bridget Oppenheimer to the grand opening of the Johannesburg Civic Theatre. In 1992, Thomas and Annie Cullinan's great-granddaughter Anne Gretchen Robinson sold the necklace to Stephen Silver, president of Stephen Silver Fine Jewelry in Menlo Park, California. It was loaned to the Smithsonian for exhibition in 1994, and then shown at Harrah's casino's sixtieth anniversary exhibition, Dazzling Diamonds, and traveled as part of the American Museum of Natural History's Nature of Diamonds show. In 2010, in recognition of its exquisite beauty and historical significance, Stephen Silver presented the Cullinan Blue Diamond Necklace to the Smithsonian National Gem Collection for all to enjoy.

Kimberley Diamond

The fancy yellow 55.08-carat Kimberley Diamond is noted for its iconic emerald-cut shape and classically elegant Baumgold Brothers necklace setting. Named for the famous Kimberley mining region in South Africa, where it was found in about 1940, the diamond is set in a platinum necklace and accented with eighty baguette-cut diamonds weighing 20 total carats. In July 2019, Bruce Stuart generously gifted the Kimberley Diamond necklace to the Smithsonian Institution.

ABOVE
The Kimberley Diamond fluoresces blue under ultraviolet light.

OPPOSITE
The 55.08-carat fancy yellow Kimberley Diamond, 31.69×17.46×9.64 millimeters (1.25×0.69×0.38 inches), was set into a necklace fabricated by Baumgold Brothers in about 1940.

Investigation into the history of the Kimberley Diamond typically reveals some variations on a story that begins with an approximately 490-carat diamond crystal, sometimes described as elongated and flat, found in the 1870s or 1880s (or in one case, between 1869 and 1871) in the famed South African Kimberley mine. Some references then suggest, without citations or evidence, the diamond became part of the Russian Crown Jewels. It supposedly reached Europe during the upheavals of the 1917 Bolshevik Revolution and was purchased by an anonymous buyer who, in 1921, had the large flat stone cut (or recut) into a 70-carat flawless modern gem. But then, as the story goes, in 1958, the stone was recut by its new owners, Baumgold Brothers Inc. of New York, to improve its proportions and increase its brilliance, and they sold it to an undisclosed collector in 1971. Various newspaper accounts of the 1960s describe the diamond as 55 or 70 carats—but we now know most of what we thought we knew about the diamond, including the preceding account, to be false.

Our research reveals that the Kimberley Diamond was, in fact, cut by the Baumgold Brothers to the 55.08-carat emerald-cut gem we know today from a 490-carat diamond found in the famous Kimberley mining region in South Africa, circa 1940. In the Diamond Information Center's 1971 edition of *Notable Diamonds of the World*, the diamond rough is referred to as the Baumgold II. The first known reference to the faceted diamond mounted in the current necklace is in a jewelry store ad in the March 10, 1940, issue of *The Miami News*.

The Kimberley Diamond traveled to many jewelry stores and other venues throughout the United States in the 1940s through the 1960s, often being featured in local newspapers and other media. In 1968, the necklace appeared on the television shows *It Takes a Thief* and *Ironside*. (The Smithsonian's Victoria-Transvaal Diamond was also cut and set into a necklace by the Baumgold Brothers, and it also appeared with the Kimberley Diamond necklace on several occasions, including the episode of *Ironside*.) The Kimberley Diamond necklace was sold to an undisclosed collector in 1971. In April 1977, chemist Dr. Herchel Smith acquired the necklace at a Sotheby's Parke-Bernet auction in New York. Dr. Smith's estate consigned the necklace to Christie's auction house in October 2002, where it was acquired by Bruce Stuart. The diamond was exhibited at the American Museum of Natural History in New York from July 2013 through June 2014.

American Golden Topaz

The light-yellow 22,892.5-carat (10.08 pounds) American Golden Topaz has 172 facets and measures 175.3×149.4×93.4 millimeters (6.9×5.9×3.7 inches). About the size of an automobile headlight, it is the heaviest faceted gem in the National Gem Collection and one of the largest in the world. So large, in fact, that the *Washington Post* newspaper noted the gift this way: "In the absurdly large gem department . . ." The massive topaz gem was presented to the Smithsonian Institution on May 4, 1988, by the rock hound hobbyists of America through the efforts of the six regional federations of the American Federation of Mineralogical Societies and Drs. Marie and Edgar Borgatta. The gem then traveled to mineral shows across the country for the next several years to the delight of thousands of hobbyists who made the gift possible.

American Golden Topaz and a 12,555-carat golden topaz sphere that has more than one thousand facets. The sphere was cut in Idar-Oberstein, Germany, from a crystal fragment that was in the Smithsonian collection.

Dr. Edgar Borgatta, sociology professor at the University of Washington, was a hobbyist gem cutter with a particular passion for topaz, and in the early 1970s he purchased a 26-pound water-worn topaz crystal that had been found in Minas Gerais, Brazil. The topaz sat for some years with other stones on his back porch. Impressed by the size and apparent transparency of the crystal, in 1975, Dr. Borgatta roughed it into a rectangular shape as a first step in creating a giant gem. The project was put on hold, however, when he realized that he did not have the heavy-duty equipment necessary to facet and polish such a massive stone.

Meanwhile, in the early 1980s, the Brazilian Princess, a 21,005-carat topaz, was introduced as the world's largest faceted gem. Dr. Borgatta, rising to the challenge, was convinced that, if cut well, his topaz could top the Princess. His search for a cutter that could handle his big stone took him to Leon Agee, a jeweler in Walla Walla, Washington. Agee adapted equipment with custom-designed counterweights to accommodate the large topaz gem. The faceting and polishing required about a thousand hours, working evenings and weekends over about two years.

But was it a record? The moment of truth arrived in January 1988 with an unofficial weighing in: Agee and Dr. Borgatta used the new electronic meat scale at the local grocery store, with claimed accuracy to 0.01 pounds, to weigh the gem. When converted from pounds to carats, the weight of the finished gem was 22,875 carats (+/- ~23 carats), more than 1,800 carats heavier than the Brazilian Princess. The official weight, determined later on a calibrated scale in Richland, Washington, was 22,892.5 carats—at the time, the heaviest faceted gem in the world. (The current heaviest faceted gem is the 31,000-carat El-Dorado Topaz at the Spanish Programa Royal Collection in Madrid.)

As the giant topaz was being finished in 1987, Dr. Borgatta approached the Smithsonian about acquiring the gem. Since the American Museum of Natural History had won the bid for the Brazilian Princess, the Smithsonian curator was especially enthusiastic about the opportunity to get an even bigger gem. (Yes, there is a friendly competition among museums.) Unfortunately, the Borgattas could not afford to donate the topaz, and the cost was beyond the Smithsonian's budget. A plan was quickly hatched, however, to partner with the American Federation of Mineralogical Societies and their six regional federations to have their members raise $40,000 needed to purchase the topaz for the Smithsonian. If successful, the Smithsonian agreed to exhibit the topaz gem at major mineral shows around the United States. The project was embraced enthusiastically, and the funds were raised in just a year, by April 1988.

The 22,892.5-carat American Golden Topaz is the heaviest faceted gem in the National Gem Collection.

Giant Topaz Crystals

Topaz is renowned for its ability to grow into huge, gem-quality crystals, but these giants—weighing 31.8 kilograms (70 pounds) and 50.4 kilograms (111 pounds)—from Minas Gerais, Brazil, are two of the world's finest. They are each extraordinary for their size, perfect crystal form, and internal transparency. What appear to be bubbles rising inside are actually imprints of albite crystals that once grew against the topaz crystal surfaces. At the time they were discovered in the 1950s, their pale color made them unattractive for cutting into gems. Today, however, crystals such as these are the essential raw material for the modern blue topaz market (exposure to radiation and subsequent heating turns pale topaz to the more prized blue hue). The crystals were gifts of Joseph M. Linsey (50.4 kilograms) and Howard G. Freeman (31.8 kilograms) in 1981.

During a trip to Los Angeles in the late 1950s, Andy Kulin of ManLabs, a metallurgical research laboratory in Cambridge, Massachusetts, saw nine large topaz crystals in the window of a store, with a sign that read "on loan from Alan Caplan, New York" at the base of the display. Kulin and his brother Phil had never seen topaz so large, but they realized the crystals would be perfect for their work in x-ray fluorescence analysis. Kulin went to New York and called every Alan Caplan in the Manhattan phone book. When he found the crystals' owner, he quickly made a deal to buy them all.

In the early 1960s, most of the crystals went to the Kulins' subsidiary company Crymet and were sliced for monochromators. However, the largest and most perfect pair was stored at ManLabs. Gene Meiran was an MIT graduate student working at ManLabs when he saw the two large crystals in the company president's office. They were scheduled to be transported to Crymet for cutting that evening. Gene was an avid mineral collector and immediately realized that the extraordinary crystals were perhaps the finest and largest ever found. According to Meiran, in remarks totally inappropriate for a lowly graduate student, he threatened to do terrible things to the president if the crystals were harmed. In any case, Meiran was able to convince the company officials that the crystals were treasures that should be preserved. The two crystals were eventually loaned to the Smithsonian Institution for exhibition, and later donated by individuals who purchased them from ManLabs. One other of the original nine crystals survived and is in the Mineralogical & Geological Museum at Harvard University.

The giant topaz crystals and the
22,892.5-carat American Golden
Topaz gem all originated from the
mines of Minas Gerais, Brazil. Large
crystals produce large gems!

Carmen Lúcia Ruby

At 23.1 carats, the Carmen Lúcia Ruby is the largest faceted ruby in the National Gem Collection and one of the finest large, faceted Burmese rubies known. The ruby is exceptional because of its size, richly saturated homogeneous red color, and outstanding transparency. It was mined from the fabled Mogok region of Burma in the 1930s. The cushion-cut stone measures 18.35×14.18×9.50 millimeters (0.72×0.56×0.37 inches) and is set into a platinum ring flanked by triangular-shape diamonds weighing 1.1 and 1.27 carats, respectively. Burmese gem rubies larger than 10 carats are extremely rare, and only a few are known larger than 20 carats. Ruby is a gem variety of the mineral corundum. Pure corundum (aluminum oxide) is colorless, but small quantities of chromium present as impurities in the corundum crystals cause the red color.

The ruby was a gift from Dr. Peter Buck in memory of his beloved wife, Carmen Lúcia.

I discovered this incredible ruby when a New York jewelry dealer brought it to my office. He said he had a special stone, but I hear those words frequently, so I was unprepared when he opened the small black box. It revealed a gem glowing the fluorescent red that is characteristic of the finest Burmese rubies. And at 23.1 carats, this stone was huge; it was a fiery blaze in my hand. I had never seen anything like it! The dealer knew we were seeking a great ruby for our National Gem Collection, and this was one of the finest and largest known—one of only a few such stones ever mined in Burma. Equally astonishing is that I had never heard about this ruby. Rare and beautiful gems such as this one are typically known and have a story. Where had it come from? According to the dealer, after the ruby was mined in Burma in the 1930s, it was owned by European families, then sold to a group of jewelers who stored it in a vault for more than a decade. The Smithsonian was one of the first places it was shown. There was no doubt it was perfect for our collection, but not surprisingly, the price was equally impressive. Historically, the iconic gems in our collection are donations. So, with great regret, I returned the ruby to its box and dutifully said to the dealer: "If you know of someone who might want to donate such a stone, I will be delighted to talk with them." I took solace in knowing I had seen one of the world's great treasures—one of the best perks of my job.

Over a year later, I received a call inquiring if we might still be interested in the ruby. The potential donor was Dr. Peter Buck, a cofounder of the Subway restaurants and a client and friend of the jeweler.

Dr. Buck's interest in donating the stone was a romantic one. He wanted to commemorate his deceased wife, Carmen Lúcia, whom he had met in New York when she was a student visiting from Brazil. Carmen had seen a picture of the ruby before she tragically passed away. Upon hearing that the Smithsonian wanted the stone for the National Gem Collection, the idea came to him of possibly donating the stone to the Smithsonian in memory of his wife.

The ruby was received in 2004 as a "gift of Dr. Peter Buck in memory of his loving wife," adding: "This gemstone is donated to the Smithsonian Institution for the enjoyment of all the people of America" and, "She would really like that people could see it and know that it was the Carmen Lúcia Ruby, and it wasn't locked away in a vault somewhere." The ruby went on exhibit that year, after it made a brief visit to Connecticut so that all the Buck family could privately experience the beauty of the special gem. During

the following years, Dr. Buck made several other major gifts to the Smithsonian that supported research, exhibitions, and education. During his visits to the National Museum of Natural History, he enjoyed standing quietly in the gem gallery and watching visitors' reactions to the beautiful ruby, and he would often say to me: "This would make Carmen so happy."

The richly saturated, homogeneous red color and exceptional transparency and clarity of the 23.1-carat Carmen Lúcia Ruby make it perhaps the finest and largest ruby on public display in the world.

Dom Pedro Aquamarine

This spectacular aquamarine obelisk towers 35 centimeters (13.75 inches) high, is 10 centimeters (4 inches) wide at the base, weighs 10,363 carats (about 4.6 pounds), and is the largest known gem aquamarine. It was fashioned in Idar-Oberstein, Germany, by acclaimed stone cutter Bernd Munsteiner. The contemporary fantasy cut features lozenge-shape negative facets stepped along the back of the obelisk to reflect and refract the light throughout the gem, giving it an ethereal glow. The giant gem stands alone as the finished piece of art, without mount or setting.

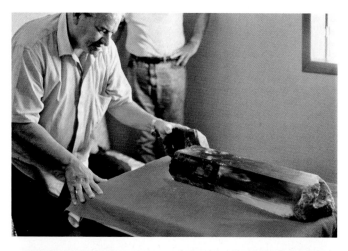

In the late 1980s, an extraordinary meter-long aquamarine crystal was found by three Brazilian miners in Pedra Azul, Minas Gerais, Brazil. It was accidentally dropped as it was removed from the mine, breaking into three pieces. The two smaller pieces, both of exceptional color, were cut and faceted into gems that were sold commercially. The finest and largest piece of the crystal weighed almost 60 pounds and was approximately 0.67 meters (2 feet) in length. This fabulous aquamarine was considered too extraordinary to simply be faceted into many smaller gemstones. Ultimately, it found its way into the hands of gem connoisseurs and experts who realized its importance and potential. In 1991, Jürgen Henn, a gem dealer from Idar-Oberstein, Germany, visited the Brazilian miner who owned the crystal, and when shown the aquamarine, he realized it was the largest, finest, most important one he had ever seen.

A year later, when Henn heard the aquamarine was for sale, he immediately sent his son, Axel, and Chico Banks, the son of Idar-Oberstein gemologist Hermann Banks, to negotiate a deal to acquire the crystal. A partnership was formed that included Jürgen Henn, the Brazilian owner, and Hermann Banks, with Bernd Munsteiner as the cutter. Axel and Chico faced many challenges transporting the valuable crystal back to Idar-Oberstein. The story is not without intrigue—including shutting down a major airport and temporarily stashing the aquamarine in the control tower, in order to securely transfer the valuable stone out of Brazil—but in the end, the crystal was safely delivered into the hands of master cutter and gem artist Bernd Munsteiner. Jürgen Henn insisted that the great crystal should be transformed into a magnificent and towering gem sculpture as unique and beautiful as the crystal itself, saying: "What nature made large, man should not make small!"

For Bernd Munsteiner, when he saw the crystal, "it was love at first sight!" The aquamarine was a beautiful transparent sea-blue color, and except for hollow tube inclusions near the top of the crystal, it had exceptional clarity. From 1992 to 1993, Munsteiner spent four months studying the crystal and then six

LEFT, ABOVE
The ~100-pound aquamarine crystal as it came from the mine in Brazil; the two smaller pieces were cut into gemstones, and the large 60-pound transparent upper section was later cut into the Dom Pedro.

LEFT
Idar-Oberstein master gem cutter Bernd Munsteiner with the completed Dom Pedro Aquamarine in the Smithsonian Gem Gallery.

months cutting, faceting, and polishing it. The finished gem was named Dom Pedro after the first two emperors of Brazil.

The Dom Pedro was first exhibited in 1993 at the annual gem fair in Basel, Switzerland. It continued to travel as an ambassador for the German government. By the late 1990s, however, the Brazilian partner wanted it sold to recoup his investment, even if that meant cutting it into smaller, readily salable gems. Fortunately, Jane Mitchell, a gem collector and friend of the Henns, had followed Dom Pedro's journey from crystal to gem. In 1996, she even helped to organize an exhibit in Palm Beach, Florida, where the Dom Pedro was displayed along with other gem carvings and sculptures to showcase the genius of the gem artists of Idar-Oberstein. When she heard the Dom Pedro might be cut up into many smaller gemstones, she and her husband, Jeffery Bland, decided to purchase the aquamarine. They continued to exhibit the Dom Pedro but eventually decided it should be in a museum, where it could be viewed and admired by many and be an inspiration to all. They generously gifted Dom Pedro to the Smithsonian Institution in 2011.

This spectacular aquamarine obelisk towers 35 centimeters (13.75 inches) high, is 10 centimeters (4 inches) wide at the base, and weighs 10,363 carats (about 4.6 pounds).

Sherman Diamond

The 8.52-carat pear-shape Sherman Diamond, 16.3×13.1×6.65 millimeters (0.64×0.51×0.26 inches), is in a pendant surrounded by fifteen round diamonds, graduated in size, and suspended from a bale with two additional diamonds. The Sherman Diamond is near colorless, with several dark inclusions.

ABOVE

This necklace was given to Sherman's daughter Minnie as a wedding gift in 1874 by Khedive Ismail Pasha, ruler of Egypt. The diamond in the necklace's central pendant is the Smithsonian's Sherman Diamond.

OPPOSITE

The 8.52-carat pear-shape Sherman Diamond is in a pendant surrounded by fifteen round diamonds.

Following General Tecumseh Sherman's success in the Civil War, he became an international celebrity. While visiting Egypt in 1873, he provided military advice to the country's ruler, Khedive Ismail Pasha. As a thank you, the khedive sent a wedding gift to Sherman's elder daughter Maria, aka "Minnie."

The "souvenir," as the khedive described it in a letter, was a fabulous diamond necklace with matching earrings. The parure arrived in New York in January 1875 and was delivered to the United States Custom House, where it was viewed by members of the Sherman family. Customs appraised the jewels at nearly $300,000 and assessed a hefty import duty. Later, the appraisal was lowered to half that amount. A *New York Times* article (January 28, 1875) reporting the arrival of the jewels caused a media frenzy, and thousands of people descended on the Custom House for a look. The crowds so annoyed the deputy customs collector that he had the jewels removed to the subtreasury vault, where they remained until they were transferred to the general in July 1876. Sherman then had them delivered to the treasury department in Washington, DC, where they were stored with his ceremonial sword.

As commanding general of the United States Army, neither Sherman nor his family was permitted to receive gifts from foreign heads of state without consent of Congress. Congress approved a joint resolution in January 1875 and then passed a bill authorizing the gift—duty-free. Because of the financial burden to the young couple—imposed by the yearly property tax on the jewels—an agreement made the general and his wife trustees of the jewels. Additionally, the diamonds would be equally divided among Sherman's four daughters, and they would stay in the family until the general's death. In late 1878, the necklace and earrings were sent to Tiffany & Co. to be reset into four pendants and four pairs of earrings, each of equal value but in different styles. Minnie was given the first choice and then each sister in order of age.

The large Sherman Diamond made its way to the Smithsonian through a rather meandering route. Minnie's sister, Eleanor Sherman Thackara, sold her necklace and earrings to her aunt, Mrs. Margaret Sherman (General Sherman's sister-in-law). Mrs. Sherman gave the gems to her adopted daughter, Mary Sherman McCallum, who gifted them to her daughter, Cecilia McCallum, on the occasion of Cecilia's marriage in 1923. Mrs. Cecilia McCallum Bolin, who lived in Washington, DC, donated the diamond pendant to the Smithsonian National Gem Collection in 1969 in memory of her mother. Comparison with historic photographs confirms that the 8.52-carat diamond was in the center pendant of the original necklace.

Thompson Diamonds

In the spring of 1957, Harry Winston Inc. cut three exquisite pear-shape, yellowish-brown "cognac"-hued gems from a 264-carat diamond crystal they had purchased the previous year from the Diamond Trading Company in Antwerp. The pendant features a 36.73-carat pear-shape fancy brown diamond and is surrounded by thirty-three pear-shape colorless diamonds that total 9.03 carats. The two smaller stones, 20.46 and 19.12 carats, were set in the matching earrings accented with seventy-two pear-shape colorless diamonds that total 10.75 carats. The dazzling ensemble of cognac and colorless diamonds was purchased while in production by Libby Moody Thompson from Galveston, Texas, who left them as a bequest to the National Gem Collection in 1990.

Libby Moody Thompson was born in 1898 in Galveston, Texas, daughter of William Lewis and Libby Rice Moody. Her father built the family fortune as founder of the American National Insurance Company and Moody National Bank, along with ranching and hotel interests. The family established the Moody Foundation for the "perpetual benefit of present and future generations of Texans." After graduating from high school in Galveston, Libby completed her education at the Holton-Arms finishing school in Washington, DC. She met her husband, Clark Wallace Thompson—originally from Wisconsin—at a dance in Galveston. They were married in 1918 in Richmond, Virginia, while Thompson was on active duty in the United States Marine Corp.

Clark Thompson retired with the rank of colonel after active duty during World War II, and in 1946 he was elected to the United States House of Representatives from the Texas ninth district. He had previously served as representative from the Texas seventh district from 1933 to 1935. Libby's obituary in a Galveston newspaper described her as a Washington "hostess with the mostest on the ball." Arriving in Washington after the 1946 election, she was credited with reviving the custom of social calls, making more than five hundred visits to wives of cabinet members, senators, congressmen, and justices during her first few months in town. A typical call would last fifteen minutes, and she averaged six per day. The Thompson home in Washington, DC, known as the "Texas Embassy," was legendary for lavish parties attended by Supreme Court justices, cabinet secretaries, and presidents (including the reclusive Harry Truman in 1952).

Libby's father passed away in 1954, leaving her an inheritance, and perhaps not coincidentally, a short time later she purchased the diamond suite. In 1979, the *Washington Post* named her one of the richest women in the United States. Mr. Thompson retired from Congress in 1967 and died in 1981. Libby devoted considerable time to a variety of nonprofit organizations until her death in 1990. Perhaps inspired by her time in Washington, DC, and/or by Harry Winston, she left the suite of cognac diamonds as a bequest to the Smithsonian National Gem Collection.

OPPOSITE
The three pear-shape cognac diamonds were cut from a single 264-carat diamond crystal.

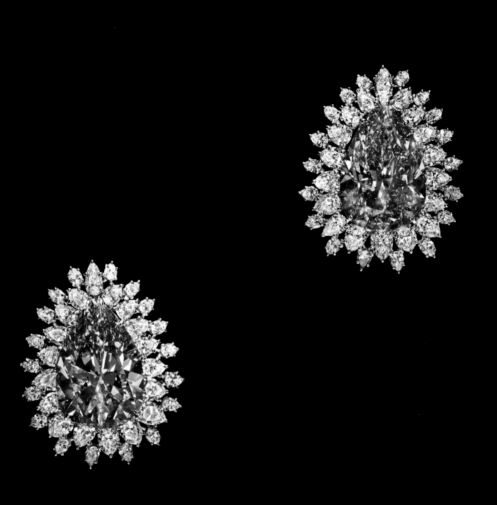

Clagett Bracelet

The Clagett Bracelet is a glittering masterpiece of craftsmanship and art deco artistry. The platinum bracelet is set with diamonds and colored stones forming a landscape with enameled figures of a hunter on horseback and another on foot hunting a lion, styled after a Persian miniature. The hundreds of individual segments are so skillfully articulated that the bracelet lies perfectly flat but also curves elegantly, as if a piece of cloth, around the wrist. Imagine fastening all the pieces of a large jigsaw puzzle, each to the other, such that the entire puzzle undulates as a single sheet. The beautiful bracelet was a gift to the National Gem Collection in 1992 from C. Thomas Clagett.

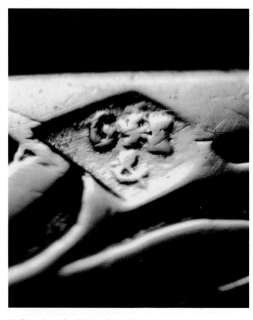

Hallmark on the Clagett Bracelet indicating it was fabricated by the Parisian firm of Geoffroy et Eisenmann. It was made during the period 1923–1925.

In the winter of 1992, we received a phone call from Washington, DC, resident C. Thomas Clagett, a retired executive for Zeigler Coal Company and later Houston Natural Gas. He explained that his wife, Nancy Leiter Clagett, had passed away several years earlier, and he was trying to decide what to do with her jewelry—particularly her favorite bracelet. Although he had agreed to sell the bracelet at auction, he was having second thoughts. He wondered if perhaps we might want it for the Smithsonian National Gem Collection. He brought the bracelet to the museum, and we were awestruck by the exquisite art deco craftsmanship in platinum, diamonds, and other gems highlighting an enameled Persian hunting scene. He couldn't remember where his wife bought it, only that it was in Europe. We assured him that we would be delighted to add this beautiful piece of jeweled art to our collection.

Our curiosity about the origin of the Clagett Bracelet launched an investigation that lasted several years. Various experts on art deco jewelry concurred that the craftsmanship and artistic style were unlike anything they had seen, with some declaring it perhaps the finest such bracelet from the period. None, however, could tell us who had made it. Our own examination revealed two hallmarks stamped on the bracelet: one shaped like a dog's head, the French symbol for platinum in the 1920s; another, the letters G and E separated by a swastika symbol, was unknown to us. We sent images of the mysterious hallmark to various experts but heard nothing.

Then, in 2009, I received news from Danusia Niklewicz of the Hallmark Research Institute. After much determined searching, she finally discovered that the unknown hallmark was from Geoffroy et Eisenmann (G & E, a jewelry maker in Paris whose symbol was a swastika) and that it had been registered on July 4, 1923. Coincidentally, at about the same time, jewelry historian Janet Zapata contacted us to share her excitement at finding a picture of our bracelet in the catalog for the International Exposition of Modern Decorative and Industrial Arts (Exposition Internationale des Arts Decoratifs et Industriels Modernes) of 1925 in Paris. The term "art deco" was derived by shortening the words "arts decoratifs" from this exhibition. The bracelet made by the workshop of Geoffroy et Eisenmann had been entered into the exposition by the jewelry firm Vever of Paris. The bracelet was one of the exposition's grand prize winners. Clearly, Thomas Clagett and his wife were correct; the bracelet was very special and deserved to be part of the National Gem Collection.

ABOVE
There are 626 old-European-cut and single-cut diamonds; round rubies form flower heads, marquise-shape leaves are made of emeralds and citrine, and a "river" of blue sapphires flows through the landscape.

BELOW & RIGHT
The art deco Clagett Bracelet was a grand-prize winner at the art deco exposition in Paris in 1925.

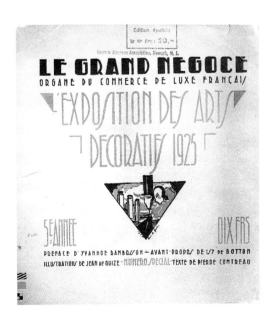

Vever (grand prix)
14, rue de la paix

Linzeler et Marchak (grand prix)
4, rue de la paix

DeYoung Red Diamond

The 5.03-carat modified brilliant-cut DeYoung Red Diamond measures 11.54×11.82×6.36 millimeters (0.45×0.47×0.25 inches) and was graded by the Gemological Institute of America as a fancy dark-reddish-brown color. Spectroscopic studies indicate that it is a type Ia diamond (i.e., containing nitrogen) and that the broadband light absorption is characteristic of natural pink and purple diamonds. The color is concentrated along planar graining in the diamond, and as with similar-colored diamonds, it is thought that the color originates from shock deformation of the diamond structure, either after formation deep in the earth or during the volcanic ascent to the surface.

During the dark days of the Great Depression, it was not uncommon for families to sell treasured heirlooms and jewelry to survive. Such was the case in Boston in the 1930s, when an employee with the jewelry firm J & SS DeYoung purchased a collection of jewelry that included a "garnet scarf pin." Later that day, a deep cleaning and more careful examination revealed that the garnets were, in fact, two red diamonds, with Mr. DeYoung dramatically announcing his discovery to his colleagues by dragging one of the stones across the store window, producing a deep scratch. When the pin was fastened, the two red diamonds, weighing 5.03 and 1.15 carats, fit together to appear as one, and considering their shapes and matching colors, were likely cut from the same original rough crystal. Recognizing their rarity, Mr. DeYoung kept his treasured red diamonds in a small cardboard box in his vault, only showing them to special visitors with an appreciative eye.

In the 1980s, while wintering at the Breakers in Palm Beach, Florida, Mr. DeYoung shared sunning time on the patio with fellow guest, and longtime Smithsonian friend, Lillian Turner, and during one of those sessions she suggested that Mr. DeYoung should add to his previous gift of the DeYoung Pink Diamond to the Smithsonian by also giving one of the red diamonds. Perhaps because of her gentle nudge, the larger red diamond was received as a bequest to the National Gem Collection from Mr. DeYoung after his death in 1986. The smaller red diamond was set into a ring for a member of the DeYoung family.

One added note: In the article "A Few Words on Fancy Colored Diamonds" in *The Jewelers' Circular* in February 1920, by Frank Wade, he relates how he had a dark brownish-red diamond that he left with a jeweler to be set into a scarf pin, and that when he retrieved it, he noted that the clerk had written on the envelope: "Set garnet in scarf pin." Might it have been the same diamond?

The 5.03-carat DeYoung Red Diamond is one of only a few diamonds known having an intense garnet-red color.

DeYoung Pink Diamond

In the late 1950s, Baumgold Brothers Inc. in New York purchased from De Beers a package of pink and blue diamond crystals. Baumgold then made a deal to sell all the finished stones from the lot to J & SS DeYoung in Boston. During the next several years, more than sixty faceted pink and blue diamonds, ranging from 0.75 to almost 4 carats, were received by DeYoung, and most were immediately sold. One 2.82-carat intense purplish-pink diamond from Tanzania was considered exceptional and was set aside by Sydney DeYoung. When Smithsonian curator George Switzer visited Boston in 1961, Mr. DeYoung showed him the pink diamond and announced his intention to donate it to the National Gem Collection. The diamond was received as a gift and placed on exhibition the following year.

Uncle Sam Diamond

In 1906, John Huddleston was surprised to find diamonds on his farm near Murfreesboro, Arkansas, but following an initial frenzy of excitement, ensuing attempts to mine the diamonds commercially were unsuccessful. The most sustained mining effort was made by the Arkansas Diamond Company between 1919 and 1925, but despite finding thousands of diamonds, the cost of recovering them was too great to turn a profit. The mine's most exciting find was made in the summer of 1924: a 40.23-carat diamond crystal that remains the largest diamond ever recovered in the United States. It was named the Uncle Sam Diamond, and the optimism around its discovery helped extend the mining operation into the next year.

The Uncle Sam Diamond was acquired from the estate of Thomas Cochran, who had been one of the largest shareholders in the Arkansas Diamond Company, by Schenck & Van Haelen of New York, who specialized in Arkansas diamonds. It was cut by Ernest G. H. Schenck to a 14.34-carat parallelogram shape and then recut to improve its brilliance to the current 12.42-carat emerald-cut gem. It is the largest faceted diamond from Arkansas.

The diamond traveled from jeweler to museum to private owner before coming to the Smithsonian. It was acquired from Ernest Schenck's estate by Boston jeweler Sydney DeYoung in 1955, then sold (around 1960) to B. Beryl Peikin of Peikin Jewelers of Fifth Avenue, New York. It was loaned to the American Museum of Natural History for exhibition the same year. Mr. Peikin passed away in 1988, and the diamond, set in a ring with two smaller Arkansas diamonds, remained with his wife until she passed away at age 102. DeYoung jewelers owned the diamond again, briefly, until Dr. Peter Buck (cofounder of the Subway restaurant chain) heard about the resurfaced gem. He immediately decided that the Uncle Sam Diamond was a national treasure that should be available for all to see, and he purchased it for the Smithsonian National Gem Collection in 2019.

Infrared spectroscopy studies by the Gemological Institute of America revealed that the Uncle Sam is a rare type IIa diamond, that is showing no indications of nitrogen impurities. Less than one percent of all diamonds found are type IIa.

Freedom Diamond

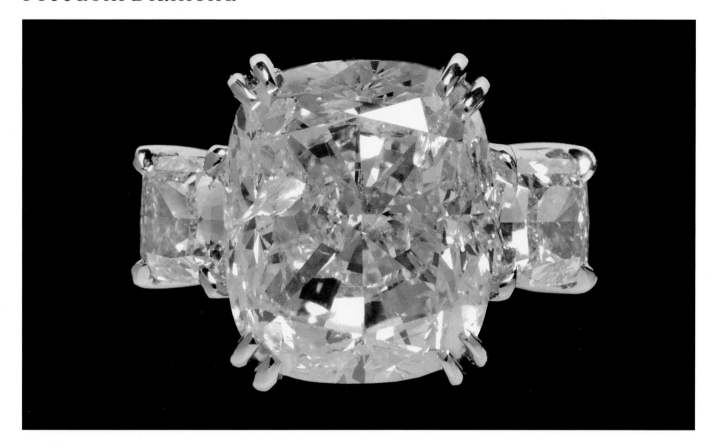

In 1975, a United States Geological Survey technician discovered a small diamond while sawing a rock from kimberlite (a volcanic rock that erupted from deep in the earth through vertical columns, called pipes) near the Colorado-Wyoming border. The discovery set off a mini diamond rush in the area. Prospectors found numerous diamondiferous pipes, but the diamonds were small and sparse. A little more than a decade later, geologist Howard Coopersmith discovered a cluster of diamond-rich kimberlite pipes in the area near Kelsey Lake, Colorado, and the company Redaurum Mines Ltd. began mining and processing operations in 1996. They produced approximately 12,000 carats of diamonds the first year and 9,000 carats the next, successfully building a local market for Colorado diamonds. Most of the diamond crystals were smaller than a carat, but in the summer of 1996, they found a 28.3-carat yellow crystal, and the following year they found a second large stone.

On July 14, 1997, a roughly octahedral-shape 28.17-carat "fairly white" crystal was handpicked from the coarse concentrate in the processing plant from Kelsey Lake No. 2 kimberlite pipe. It was sold to the Colorado Diamonds company, and they sent it to Y&M Diamond Trading Company Inc. in New York for cutting. Three weeks later, the finished light-yellow 16.87-carat cushion-cut gem became, and still is, the largest faceted diamond ever produced from a United States source. The diamond changed hands a few more times after that: Thomas Hunn Company Inc. in Grand Junction, Colorado, purchased it in 1998, then sold it to Robert Mau in 2014. Mau had it set into a platinum ring flanked by 1.5- and 1.53-carat diamonds, kept it for a time, then he and his wife, Kathy, generously gifted it to the Smithsonian National Gem Collection in 2019 as the Freedom Diamond.

As for the mine where it originated, it eventually failed. Despite considerable proven diamond reserves still in the ground, all mining at Kelsey Lake ceased in 2001, and the area was completely reclaimed between 2003 and 2006 to its original natural setting.

Smithsonian Cup

The Smithsonian Cup was fashioned from a piece of agate that caught the eye of Gianmaria Buccellati at a gem show. After it sat on his desk for several years, the patterns in the stone inspired the cup. The Smithsonian Cup's main elements include yellow, white, and rose gold, as well as hints of silver and forty medallions in mother-of-pearl. The cup was twenty-four years in the making as Buccellati contemplated how to create a chalice that harmonized the natural elements of the agate stone with the ornate decorations handcrafted by House of Buccellati goldsmiths. The cup stands 12.5 centimeters (5 inches) tall. Gift of Gianmaria Buccellati in 2000.

Dunn Pearl Necklace

This elegant necklace, made by Cartier in London in 1928, has five rows of graduated natural, round Persian Gulf pearls with a magnificent diamond, pearl, and platinum clasp. The 339 pearls are cream color and well matched, and range in size from 3.3 to 7.8 millimeters (0.13 to 0.31 inches). The art deco platinum clasp has six diamond chains on each side that attach to the strands of pearls. There are 428 old-mine-cut diamonds in the clasp and chains, with a total weight of approximately 16 carats. The necklace was sold in 1929 to Alice Marie Croll, wife of William Luther Croll, an American dentist who practiced in London. The necklace was gifted to her niece Virginia McKenney Dunn (wife of Arthur Wallace Dunn) of Washington, DC, who had some pearls removed to shorten the strands. Mrs. Dunn donated the necklace to the Smithsonian National Gem Collection in 1977.

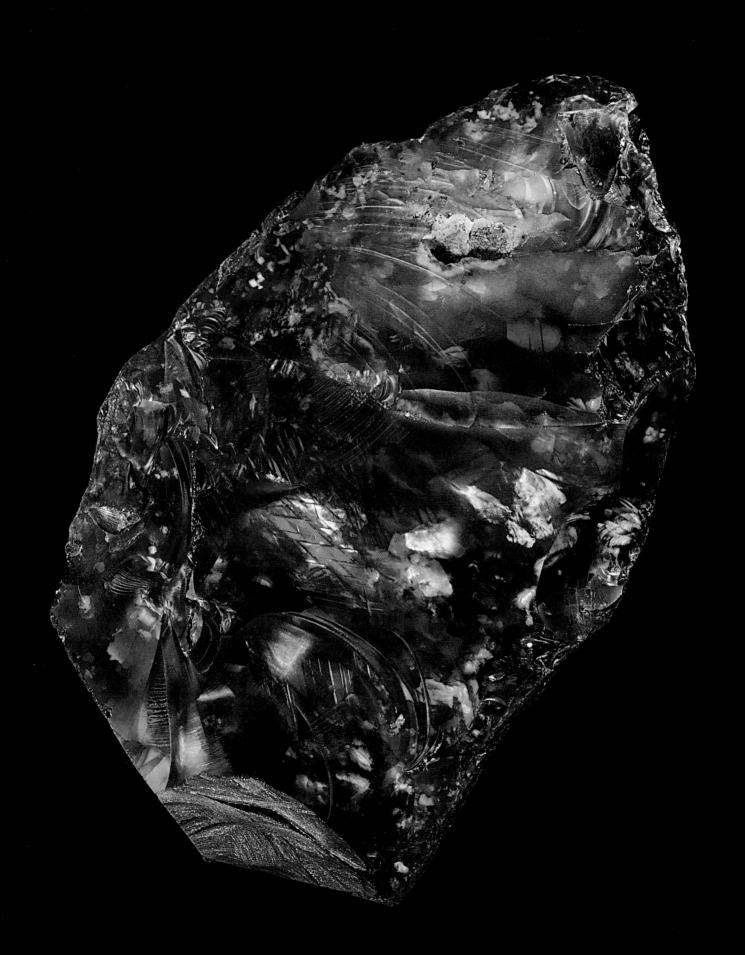

A Dozen Notable Gems

The gems here represent some of the wonderful diversity of the earth's minerals represented in the Smithsonian National Gem Collection. Many of these gems are among the finest examples known—because of their exceptional sizes, including several world-record holders, or because they are minerals rarely cut as gems, or simply because they are stunningly beautiful.

Maharani Cat's Eye

The 58.19-carat Maharani chrysoberyl from Sri Lanka is prized for its rich honey-yellow color, exceptional size, and well-defined chatoyancy, or "cat's eye." The gem was acquired in 1961 from Noel Levko of the Gem Radiation Laboratory in New York, in exchange for a number of small faceted diamonds that had been transferred to the Smithsonian from the United States Customs office.

The optical phenomenon of chatoyancy can be found in many gemstones, but the most popular and highly prized example is that of the mineral chrysoberyl (beryllium aluminum oxide). In fact, the term "cat's eye" is synonymous with chrysoberyl, owing to the resemblance of the phenomenon with the vertical slits of a cat's pupils. The "eye" that the stone displays when it is cut en cabochon (a rounded domed shape) in the correct orientation is caused by light reflecting off inclusions—fine, needlelike crystals inside the stone, commonly of the mineral rutile.

Picasso Kunzite Necklace

This spectacular necklace was designed by Paloma Picasso and made by Tiffany & Co. in 1986 to celebrate the 150th anniversary of Tiffany and as an homage to George F. Kunz, noted Tiffany gemologist. It features a 396.3-carat cushion-cut kunzite gem from Afghanistan wrapped by 18-karat yellow gold ribbons of pavé diamonds. The gem is set in platinum and suspended from a necklace of thirty South Sea baroque pearls.

Paloma Picasso is a designer best known for her jewelry designs for Tiffany & Co. She is the daughter of artist Pablo Picasso and painter Françoise Gilot. According to Ms. Picasso: "The kunzite necklace is one of the jewelry designs I am most proud of."

Kunzite is a gem variety of the mineral spodumene that was first found in Pala, California, in 1902 and was named after George F. Kunz. Kunzite gemstones are typically shades of violet and pink, caused by impurities of manganese. Gift of Tiffany & Co. in 1989.

Eternal Flame Opal

The 568-carat Eternal Flame Opal is the largest and most significant black crystal opal known from Australia's Tintenbar opal field in New South Wales. Tintenbar opals are unique because, unlike other Australian sources, they formed in seams and pockets of volcanic rock. The gem's glassy transparency reveals the fiery play of color from deep within the stone. The opal was mined in the 1970s by Campbell Bridges. Gift from the Bridges family in 2019 in loving memory of Campbell Bridges.

Peridot

More than three thousand years ago, Egyptians fashioned beads from golden-green crystals mined on an island in the Red Sea. Known to the Greeks and Romans as Topazios, this island off the coast of Egypt was one of the most important sources for fine peridot, the gem variety of the olivine-family mineral forsterite. Originally called topazion, after the island, this gem was renamed peridot in the eighteenth century. The island is known today as Zabargad, the Arabic name for peridot. Other major sources of peridot include Burma, the U.S. (Arizona), Norway, Brazil, China, Australia, and Pakistan. Peridot is a magnesium-iron silicate; pure forsterite is colorless, but iron atoms replacing some of the magnesium produce the green shades.

Five continents are represented in this impressive collection of peridot gems. The 311.8-carat gem in the center is from Zabargad, Egypt, and is the largest faceted peridot known. It was part of the original Roebling Collection donated to the Smithsonian in 1926, and supposedly it adorned an image of a saint in an Austrian church for several centuries. The gem in the necklace (34.65 carats) is from Arizona and is the largest peridot known from the United States. Other peridot gems, clockwise from the center, are: from either Egypt or Burma (103.2 carats); from Burma (round, 286.6 carats); from Norway (4.08 carats); a rare gem from McMurdo Sound, Antarctica (3.07 carats); another gem from Arizona (8.93 carats); from Pakistan (18.3 carats); and a 122.7-carat cushion-cut gem from either Burma or Egypt.

This smoky-citrine quartz gem, dubbed the "football," weighs a whopping 19,747 carats (8.69 pounds) and was cut in 1987 by American gem cutter Michael Gray from a crystal found in Brazil. It measures 25.5×14.1×10 centimeters (10×5.5×4 inches).

LEFT
Smoky-citrine quartz crystals from Minas Gerais, Brazil, weighing 53 kilograms (117 pounds). Gift of the Independent Jewelers Organization.

Petersen Tanzanite Brooch

The gem variety of zoisite known as tanzanite was first discovered in 1967 in the foothills of Mount Kilimanjaro in Tanzania, which is still the only known source. This Harry Winston Inc. brooch showcases two matched tanzanite gems, 30 carats total, in a platinum setting with accenting diamonds. The deep violet-blue tanzanite "flowers" can be removed and worn as earrings. Gift of Donald and Jo Anne Petersen in 2002.

Baryte

At 714.67 carats, this is the largest faceted baryte gem known. The original crystal was found in Brazil, and the stone was cut by Erwin Poes in Idar-Oberstein, Germany. Although a spectacular display gem and demonstration of the gem cutter's skill, baryte (barium sulfate) is too fragile to be used in jewelry.

Demantoid Garnet

Demantoid is the green gem variety of the garnet-family mineral andradite (calcium iron silicate). It is the rarest and most valuable garnet gem. Demantoid was first discovered in Russia's Ural Mountains in 1851 and was popular in Carl Fabergé's jewelry and precious objects made for the Russian tsars. Demantoid means "diamond-like," as the gem displays an adamantine luster with great brilliance and fire. Gems larger than a few carats are rare. At 11.24 carats with intense emerald-green color, this cushion-cut stone, found in Russia in the late 1990s, is one of the largest and finest faceted demantoid gems known. Gift of Smithsonian Gem and Mineral Collectors in 2011.

Dibble Fluorite

This mammoth 3,965.35-carat gem, nicknamed Big Blue, is one of the largest faceted fluorites known. It was cut by American gem cutter Arthur Grant from a crystal recovered from the Minerva No. 2 mine in Illinois. Because fluorite (calcium fluoride) is relatively soft and fragile, it is rarely used in jewelry, but its many attractive colors make it a popular collector's gem. Gift of Harold and Doris Dibble in 1992.

Dibble Calcite

Calcite (calcium carbonate) is a common and widespread mineral with many crystal forms and colors. Limestone and marble are composed of tiny calcite crystals. Calcite gems are too soft and fragile to be worn in jewelry. This calcite gem from St. Joe No. 2 mine in Balmat, New York, is a spectacular display piece and a testament to the gem cutter's art. At 1,865 carats, it is one of the largest and finest faceted calcite gems known. It was faceted by Arthur Grant. Gift of Harold and Doris Dibble in 1994.

Mogul Emerald Necklace

This emerald's long and circuitous journey to the Smithsonian took over three hundred years. The original crystal was mined in Colombia and shipped by the Spanish to Europe in the 1600s. It was sold in India, where it was carved with a floral motif for the Mogul rulers in about 1700 (note the curved drill holes on the left and right edges of the gem that were used to attach it to a cloak or turban). The rounded six-sided shape of the gem indicates that it was a cross-sectional slice cut through the original emerald crystal. It was later put in the diamond setting in India, and the top diamond and platinum ornament was added in Paris in the 1920s.

It was purchased as a brooch mounted on a gold stick by Edith Taylor Huntington, and later the platinum and diamond chain was added. The necklace was passed down to her daughter Madeleine Murdock of New Jersey, who left it as a bequest in 2007 to the National Gem Collection.

Blue Flame Lapis Lazuli

Lapis lazuli has been valued for its intense blue color for thousands of years. The ancient Egyptians fashioned it into beads and carvings and powdered it for use as eye shadow. Historically, all gem lapis lazuli was produced from a remote valley high in the Hindu Kush mountains of Afghanistan. Lapis lazuli is a rock composed of several minerals, primarily blue lazurite, with minor amounts of white calcite and golden pyrite. Lazurite is a complex sodium calcium aluminum silicate with variable amounts of sulfur and chlorine. The deep-blue color is caused by light interacting with the sulfur atoms. The name is derived from the Persian word *lazhward*, meaning "blue." This striking 75-kilogram (165-pound) free-form lapis lazuli sculpture was fashioned in Idar-Oberstein, Germany. Gift of Jane Mitchell and Jeffery Bland in 2015.

Gem Families

These various gem families showcase the rich breadth and depth of the Smithsonian National Gem Collection. They highlight more spectacular diamonds, emeralds, and sapphires, but also introduce magnificently colorful gems cut from quartz, tourmaline, spodumene, garnet, spinel, and other minerals. Some are in jewelry, and others exemplify the simple elegance of a skillfully faceted gem. Many rank among the finest of their kind, and each has a story.

Diamond

Most diamonds formed two to three billion years ago more than a hundred miles deep inside the earth, where temperatures exceed 1,200 degrees Celsius (2,192 degrees Fahrenheit). The tremendous pressure of the overlying rock compresses carbon atoms into an exceptionally strong crystal structure, with each carbon atom tightly bonded to four other carbon atoms. Some diamonds were carried to the earth's surface in the magma of a deep-seated volcanic eruption, where after tens to hundreds of millions of years, those found by miners became the gems we prize today. A diamond's superior hardness and the dazzling brilliance and fire of a well-cut stone have earned it the title of "king of gems."

ABOVE
Oppenheimer Diamond
The 253.7-carat (50.74-gram)
Oppenheimer Diamond is in the shape
of an octahedron (an eight-sided
double pyramid), which is a common
shape for diamond crystals. This
diamond is 3.8 centimeters (1.5 inches)
in height and was discovered at the
Dutoitspan mine near Kimberley,
South Africa, in 1964. The yellow color
is due to impurities of nitrogen that
replaced some of the carbon atoms as
the crystal formed. The Oppenheimer
Diamond is unusual because
diamonds of this size are rarely left
uncut. It was a gift in 1964 from Harry
Winston in memory of Sir Ernest
Oppenheimer, former chairman of
the board of directors of De Beers
Consolidated Mines from 1929 until
his death in 1957.

RIGHT
Wilkinson Diamond Brooch
This brooch displays a dazzling array
of seventy-one orange to brown fancy
color diamonds. The marquise-cut,
pear-shape, and round brilliant-cut
diamonds range from 0.3 to 2.5 carats
and total 61.3 carats. Gift of Leonard
and Victoria Wilkinson in 1977.

Marquise Diamond Ring

This stunning marquise-cut 28.29-carat diamond from South Africa measures 36×16.9×8 millimeters (1.4×0.67×0.31 inches). It was set by Cartier for Marjorie Merriweather Post in the 1920s or 1930s in a platinum ring surrounded by four trilliant, eight baguette, and sixty-four round brilliant diamonds. The popular claim that the marquise cut was inspired by the shape of a lady's lips is almost certainly a romanticized and apocryphal explanation for the diamond's shape. Gift of Mrs. Adelaide C. Riggs, Marjorie Merriweather Post's daughter, in 1964.

Lacloche Diamond Bracelet

This exquisite and elegant art deco diamond and platinum bracelet was fabricated in the late 1920s by the acclaimed Lacloche family jewelers of Paris and Madrid. It features an intricate pattern of marquise, baguette, square, and round-cut diamonds. Bequest from Margaret McCormack Sokol in 2007.

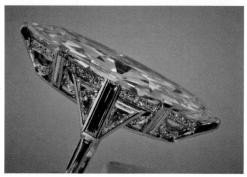

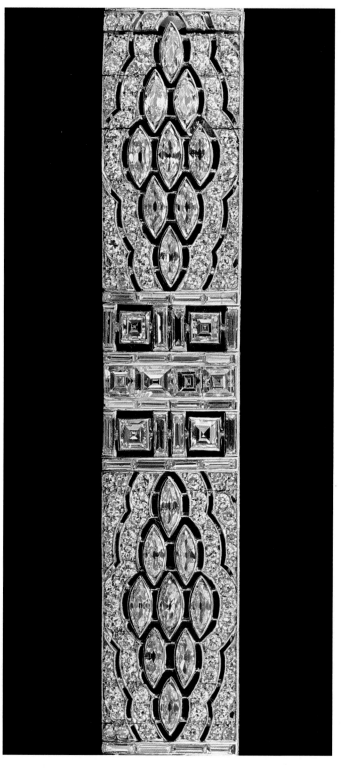

BELOW
Argyle Cognac Diamond
During operations from 1982 until closing in 2020, the Argyle mine in Western Australia was one of the world's largest producers of diamonds by volume, and famous as a source of rare pink diamonds. The bulk of the mined diamonds, however, were brown (marketed as "cognac," "champagne," or "chocolate") and smaller than a carat. This glittering 2.09-carat, cognac-colored diamond was presented to the Smithsonian National Gem Collection in 2005 by Janette Howard, wife of Australia's Prime Minister John Howard, as a symbol of the friendship and mutual respect between Australia and the United States on behalf of Sydney jeweler Nicola Cerrone and Rio Tinto's Argyle diamond mine.

TOP
Royal Butterfly Brooch
The intricate design and craftsmanship of the Royal Butterfly Brooch highlights 2,318 gems totaling 77 carats. Four large faceted diamond slices stacked atop a pavé layer of faceted diamonds form the centerpieces of the wings. The Royal Butterfly is set with sapphires and diamonds (and rubies and tsavorite garnets on the reverse side). Illumination by ultraviolet light (above) displays a hidden beauty in the Royal Butterfly, revealing the dramatic fluorescence for many of the diamonds (yellow, blue, and green) and sapphires (red and yellow). Gift of the designer Cindy Chao in 2010.

Corundum: Sapphire and Ruby

Rubies and sapphires are gem varieties of the mineral corundum. Corundum is aluminum oxide, and is colorless when pure. Fortunately, the earth is a rather dirty place to grow crystals, and most have some trace impurities that impart the brilliant colors that transform ordinary corundum into sapphires and rubies. A smidgeon of chromium replacing some of the aluminum in a corundum crystal ignites the brilliant red of ruby, but traces of iron and titanium create a sapphire's cool blue. Nature's various concoctions of a bit of this and that produce the delightful range of hues in fancy-colored sapphires. Its great hardness, second only to diamond, and unsurpassed color palette make corundum gems some of the earth's most prized treasures.

OPPOSITE
Sapphires
Although typically thought of as blue, sapphires exhibit a variety of colors, as seen here. The sapphires range from 10.3 to 92.6 carats; the green and large yellow gems are from Burma, and the others are from Sri Lanka.

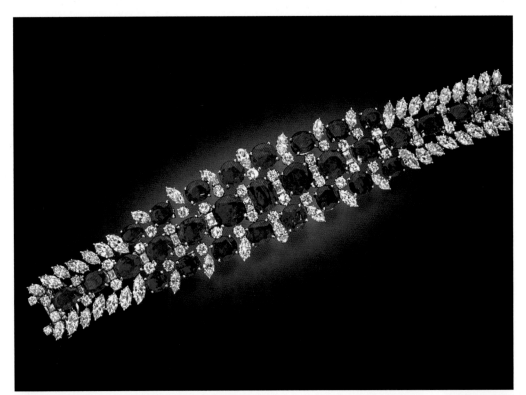

Winston Ruby bracelet

This platinum bracelet by Harry Winston Inc. has thirty-one Burmese rubies totaling 60 carats, accented with 27 carats of diamonds. It was made in 1950 utilizing the owner's old rubies. Anonymous gift in 1961.

Star of Katandru

The 16.21-carat Sri Lankan star ruby, the Star of Katandru, has an exceptionally strong and well-defined star. It was named for the children of the donor, Katherine and Andrew. Gift of Jeffrey Bilgore in memory of I. George Heyman in 2004.

Sapphire and Diamond Brooch

The sapphires, 2.78 carats total weight, in this late Victorian-era flower brooch are from Cambodia and are set with eighty rose-cut diamonds in silver and gold. The brooch was likely made in England in the 1880s. Gift of John F. Barnard in 2002.

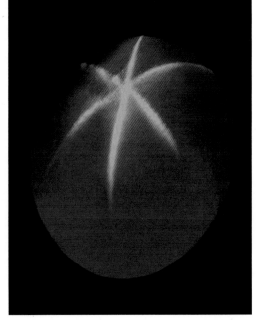

LEFT
A 175.1-carat sapphire from Sri Lanka shows the typical shape of sapphire crystals. The color variation reflects changes in composition as the crystal grew.

BELOW
Yogo Flower Brooch

This elegant wild-rose flower brooch dates from the late 1800s and features over 50 carats of sapphires from Yogo Gulch, Montana, set with diamonds. Yogo sapphires have been mined sporadically since 1895 and are greatly valued for their uniform clarity and rich cornflower-blue color. Gift of Dr. Coralyn Wright Whitney in 2020.

FAR LEFT
Conchita Montana Sapphire Butterfly

The Conchita Montana Sapphire Butterfly highlights the dazzling color range of sapphires from Montana. The 18-karat yellow-gold pendant/brooch is set with 27.97 carats of sapphires, most from the famed Rock Creek area. The brooch was designed by Paula Crevoshay using sapphires donated by Fine Gems International. Gift of Paula Crevoshay and Robert Kane in 2007.

LEFT
Star Sapphire Pendant

This Edwardian-style (1901–1915) platinum necklace designed by American jewelry firm Marcus & Co. features a 60-carat sky-blue star sapphire from Sri Lanka in a setting studded with 126 diamonds accented with bows. Gift of Mrs. W. C. Crane in 1981.

Beryl: Emerald, Aquamarine, Heliodor, and Morganite

Emeralds and aquamarines are gem varieties of the mineral beryl. Beryl is beryllium aluminum oxide and is colorless when pure. The many colors of beryl gems are caused by a variety of impurity atoms that were incorporated into the crystals as they grew. A few atoms per thousand of chromium make beryl crystals emerald green, and iron gives rise to the rich ocean-blue hues of aquamarine and makes heliodor golden-yellow. Morganite is beryl colored pink by impurities of manganese.

LEFT

Morganite gems ranging from 56 to 331 carats show the range of hues for this pink variety of beryl. The purple-pink gem in the front left and the deep pink gem at right are from Madagascar; the others are from Brazil. Morganite was named by gemologist George F. Kunz after his patron, financier J. P. Morgan.

OPPOSITE

Impurities of iron in different chemical states are responsible for the colors in these impressive aquamarines and green beryl gems from Brazil, ranging from 911 to 2,054 carats. The largest gem was cut by John Sinkankas from a crystal that was part of the Washington Roebling collection.

ABOVE, LEFT

This 147.23-carat golden heliodor is from Minas Gerais, Brazil. Gift of Patrick Coughlin in 2014.

ABOVE, RIGHT

The yellow glow of heliodor is due to impurities of iron. Heliodor gets its name from two Greek words meaning "sun" and "gift." This 275.31-carat gem is from the Ukraine. Gift of Tiffany & Co. Foundation in 2016.

LEFT

This supersize morganite gem, 1,377 carats, is about one-half the size of a baseball and is one of the largest known. It was faceted by American cutter Buzz Gray from a crystal mined in Minas Gerais, Brazil. Gift of the Smithsonian Gem and Mineral Collectors and the Winston Foundation in 2020.

Gachala Emerald

The 858-carat Gachala Emerald crystal, five centimeters (two inches) tall, was found at the Vega de San Juan mine in Gachala, Colombia, in 1967. It was acquired by gem cutter Al Horn in New York. When visiting the Smithsonian National Museum of Natural History to reacquaint himself with the Gachala Emerald in the late 1990s, Horn related how, on the day he planned to start cutting the crystal, he received a phone call from Harry Winston saying he would like to see the new emerald: "Winston knew every major gem arriving in New York."

He took the crystal to Winston's office, where Winston explained that he would like to purchase the crystal for the Smithsonian collection. Winston knew exactly how much Horn had paid for the emerald and offered a fair price to complete the deal. Rarely are emerald crystals of such size and superb color preserved; they are usually cut into gems—as this one would have been if not for that timely phone call. Harry Winston donated the Gachala Emerald to the Smithsonian in 1969.

Most Precious Aquamarine

Aquamarine, as the name suggests, is the sea-blue variety of the mineral beryl. Its color depends on the relative amounts of impurities of iron in two different chemical states in the stone. The serenely beautiful 1,000-carat fancy rectangular-cut Most Precious aquamarine gem is prized for its size, rich color, and exceptional clarity. It is named for the perfume Most Precious, which was introduced by Evyan Perfumes in 1951. Evyan owner Dr. Walter Langer contacted Smithsonian curator George Switzer in 1961 to say that he was purchasing the aquamarine gem, which originated from Morambaya, Minas Gerais, Brazil, for $10,000 and planned to use it as a sales promotion. When the gem was no longer needed, he offered to donate it to the Smithsonian. Switzer saw the aquamarine in New York later that year and concluded that it would be a "most welcome addition to the National Gem Collection." Numerous newspaper accounts chronicled the gem's journey to major cities around the country from 1962 to 1963. The Most Precious aquamarine was formally presented to the Smithsonian in September 1963 as a gift of Dr. W. Langer and Evyan Perfumes Inc. The gem is paired here with an aquamarine crystal, also from Minas Gerais, Brazil, weighing 15,256 carats (3.1 kg or 6.7 pounds).

ABOVE

The dark lines in these cabochon emeralds (5.5 and 6.9 carats) from Colombia are carbonaceous material that was trapped and oriented by the growing crystals. Such gems are called trapiche emeralds, after the Spanish word for "gears" (used in crushing sugarcane), which the pattern of inclusions resembles.

LEFT

This 1,401-carat aquamarine from Brazil was cut in the 1930s and is extraordinary for its size, perfect clarity, and classic sea-blue color. It was abandoned in a bank deposit box and eventually sold to a collector before being donated to the Smithsonian by the Bhupendra Mookim family in 2018.

Tourmaline

The tourmaline family consists of more than fourteen distinct minerals, but only one—elbaite—accounts for most of the tourmaline gemstones. Although best known in shades of green and red, elbaite can also be blue, purple, yellow, or colorless. Moreover, single crystals of elbaite can show several colors, either along their lengths or from the inside out, making it possible to cut unique multicolored gems. This color zoning results from different types of atoms that were incorporated into the growing crystal.

LEFT
The Candelabra elbaite tourmaline is from the Tourmaline Queen mine near Pala, California. The elbaite crystals sit on a quartz and albite base. The middle crystal is 12 centimeters (4.7 inches) tall.

BELOW
An elegant carving in elbaite tourmaline from Mozambique stands 5.3 centimeters (2.1 inches) high. The two orientations show the dramatic pleochroism in many tourmaline gems—the two colors resulting from light being absorbed differently as it passes through the stone in different directions.

ABOVE
This exquisite suite of three elbaite gemstones was cut from color-zoned crystals mined in Mozambique. The gems weigh 60.73, 75.24, and 90.03 carats. Gift of Somewhere in the Rainbow in 2009.

LEFT
Since their discovery in the 1980s, copper-bearing elbaite tourmaline from Paraíba, Brazil, have been highly prized for their neon-blue and -green colors, and still today, Paraíba tourmalines are among the most coveted and expensive colored gemstones. The gems here range from 1.22 to 6.69 carats. The ring was a gift from Doug and Donna Strom in 2018.

BELOW
Elbaite tourmaline crystals in quartz from Paraiba, Brazil.

RIGHT
In the summer of 2013, a pocket of gem elbaite tourmaline was uncovered at the Havey Quarry in Maine. The miners recovered this 9-centimeter (3.5-inch) crystal (and matching section found separately in the pocket) and a smaller gem-quality crystal, which was faceted by Larry Woods into this modified rectangular-cut gem weighing 30.62 carats. Maine is noted for rich teal-colored tourmaline and was one of the earliest gem mining regions in the United States. Gift of Frances M. Seay in 2014.

Quartz

Quartz is one of the most abundant minerals in the earth's crust. It is the major constituent of beach sand, but it's also found as spectacular crystals that can be cut into gems. Pure quartz is silicon dioxide and is colorless, but tiny amounts of impurity atoms give rise to a range of vivid colors. Iron and natural radioactivity in surrounding rocks make amethyst purple, and trace amounts of aluminum and radiation produce smoky quartz. Fine-grained varieties, such as agate chalcedony and jasper, which are composed of tiny fibrous or granular crystals, are prized for their wondrous profusion of colors, textures, and patterns.

OPPOSITE
This assortment of quartz gems from various localities includes amethyst (purple), citrine (golden-yellow), rose (pink), smoky (dark-brown), and rock crystal (colorless). The largest gem, the shield-shape citrine (top right), weighs 636 carats.

LEFT

Actress Angelina Jolie models the 18-karat yellow gold Jolie Citrine Necklace, which features sixty-four graduated bezel-set cushion-cut Brazilian citrine gems highlighted by a 177.11-carat pear-shape citrine drop. Citrine is the golden-yellow to orange variety of quartz, colored by impurities of iron. The name comes from the French *citron*, meaning "lemon," in reference to its color. Gift of Angelina Jolie and Robert Procop in 2015.

BELOW

A 56-carat Siberian amethyst is featured in this classic Louis Comfort Tiffany gold necklace (circa 1915). The naturalistic and decorative style of the pendant, featuring vines, leaves, and grapes, is characteristic of art nouveau jewelry from the turn of the twentieth century. The pendant is suspended from an 18-karat yellow gold chain that has a sinuous double figure eight design. Gift of June Rosner and Russell Bilgore in 2007.

OPPOSITE

The 7,478-carat quartz egg from Brazil sparkles with 240 facets and was fashioned by American gem cutter John Sinkankas.

145

RIGHT

The intense reddish-purple hue of this 78.3-carat trilliant-cut amethyst gem—faceted by John Dyer—is characteristic of stones from a recent discovery in Rwanda. Gift of Smithsonian Gem and Mineral Collectors in 2017.

BELOW, LEFT

Light scattered off closely spaced layers of quartz crystals produces the iridescent colors of this iris agate from Oregon.

BELOW, RIGHT

Agate, with its dramatic concentric color bands of fine-grained quartz, is a popular gem material for jewelry and objets d'art. This cut and polished agate is from Mexico.

LEFT
A halo of rose quartz crystals grew on previously formed smoky quartz crystals in this specimen from the Sapucaia mine, Minas Gerais, Brazil (8 inches tall).

BOTTOM, LEFT
The 172.23-carat fancy-cut gem was cut by Ryan Anderson from amethyst found at Hallelujah Junction, Nevada.

BELOW
The exquisite 214.15-carat ametrine (natural combination of amethyst and citrine) from the Anahi mine in Bolivia was carved and donated by gem artist Michael Dyber in 2005.

Garnet

The name "garnet" refers to a family of fifteen similar, but distinct, silicate minerals, five of which are commonly used as gems (almandine, pyrope, grossular, spessartine, and andradite). The color of a garnet is determined by its composition and can range from the more well-known red to various shades of purple, orange, and yellow to vibrant green. Garnets were used as gems by the early Egyptians and prized by the Greeks and Romans. The name comes from the Latin *granatum*, meaning "seed-like," because small garnet crystals resembled reddish-colored pomegranate seeds.

The pyrope garnets in this antique hairpin are from the Czech Republic, historically the primary source of garnets that were immensely popular in Victorian jewelry.

ABOVE

This colorful garnet assortment highlights an almandine from Umba Valley, Tanzania (center), and (clockwise from top) almandine (rhodolite) from Umba Valley, spessartine from Nigeria, grossular from Mali, grossular (tsavorite) from Kenya, pyrope from Umba Valley, and grossular from Sri Lanka. The gems range from 3.1 to 34.1 carats.

RIGHT

One of the rarest and most prized garnet gems is the brilliant green variety of grossular garnet called tsavorite, named for the deposit discovered in Kenya's Tsavo National Park in 1970. The color is due to the presence of vanadium and chromium in the garnet structure. The Kenya-Tanzania border region remains the only major tsavorite source. This exceptional 15.93-carat tsavorite gem is from the Scorpion Mine in Kenya. Gift of the Tiffany & Co. Foundation in 2006.

Zircon

Zircon is zirconium silicate and is commonly found as water-worn pebbles in gravel deposits in Thailand, Cambodia, Vietnam, Tanzania, and Sri Lanka. They are typically brown, reddish-brown, green, or yellow. Most zircon gemstones are heat-treated to enhance their colors; shades of blue and golden-orange or red are most popular. Zircon's brilliance and fire, or dispersion (ability to separate light into a rainbow of color), are near to that of diamond.

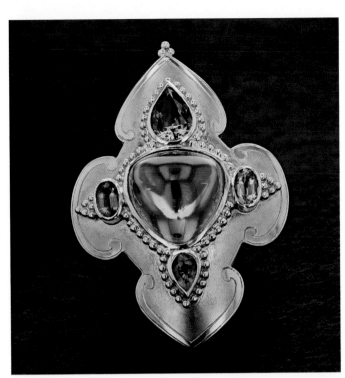

OPPOSITE

This 34.82-carat natural color zircon gem was recently mined in the Tanga Province in northeastern Tanzania.

LEFT

A natural green 43.79-carat zircon from Sri Lanka is the centerpiece of the George Pendant, named after George Crevoshay, who cut the cabochon gem. It is set in gold surrounded by four green elbaite gems from Maine, California, and Africa. The pendant was designed and donated by Paula Crevoshay in 2005.

BELOW

The zircon gems here are from Thailand (103.2 carats, blue, and 106.1 carats, orange) and Sri Lanka (48.3 carats, colorless, and 97.6 carats, green).

Spinel

Spinel gems are often mistaken for their more famous corundum cousin—the ruby. Many of the world's most famous "rubies" are in fact red spinels, including the Black Prince "Ruby" in the British Imperial Crown. Contributing to the confusion is the fact that spinel and corundum have similar chemical compositions—corundum is aluminum oxide, and spinel is magnesium aluminum oxide—and they are commonly found in the same geological deposits. Pure spinel is colorless, but as with corundum, trace amounts of chromium, iron, and other elements in spinel crystals produce a stunning array of colorful gems.

This gem-quality octahedral spinel crystal in white marble is from Mogok, Burma.

RIGHT
High luster and perfect octahedral spinel crystals found in the Mogok region of Burma are called *Anyon nat thwe*—spinels that have been cut and polished by the spirits.

BELOW
In 2007, several large spinel crystals were found near the town of Mahenge in Tanzania. One was over a hundred pounds and yielded thousands of carats of faceted gems prized for their bright watermelon pink-red color, including this 10.2-carat stone in a ring surrounded by diamonds. Gift of Smithsonian Gem and Mineral Collectors in 2019.

BOTTOM
Sri Lanka is a traditional source of great spinel gems. The rich blue color of this spectacular 14.02-carat stone is caused by trace impurities of cobalt. Gift of the Tiffany & Co. Foundation in 2009.

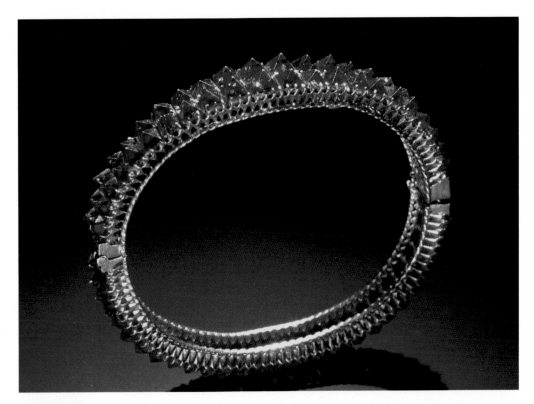

ABOVE
The Kuh-i-Lal mines in Tajikistan are one of the oldest known sources of spinel gems, and are still producing stones, such as this magnificent 40.25-carat spinel. Gift of Smithsonian Gem and Mineral Collectors in 2019.

Spodumene: Kunzite and Hiddenite

Spodumene is mined as an important ore of lithium, but it is sometimes found as lustrous transparent pink or green crystals that can be cut into gems known as kunzite and hiddenite, respectively. Kunzite was first discovered in Pala, California, in 1902 and was named for the American gemologist George F. Kunz. Kunzite gemstones are shades of violet and pink, caused by trace impurities of manganese. The most important sources of kunzite are Brazil, Afghanistan, California, and Nigeria. Vibrant green hiddenite gems mostly are found in North Carolina, and stones larger than a couple of carats are rare.

This magnificent 640-carat lavender "lotus-cut" kunzite gem from Nigeria was fashioned by gem cutter Victor Tuzlukov.

LEFT
This 880-carat heart-shape gem from Brazil is the largest faceted kunzite in the National Gem Collection.

BOTTOM, LEFT
The greenish-yellow color of this 127.08-carat gem from Afghanistan is unusual for spodumene. Gift of Dudley Blauwet in 2019.

BOTTOM, RIGHT
The Oceanview mine in Southern California produced some of the deepest-purple kunzite crystals ever found. This 164.11-carat kunzite gem was cut from a crystal that was part of the Big Kahuna pocket discovered in 2010.

BELOW
In 1897, a vibrant green variety of spodumene was found in North Carolina and was named hiddenite, after its discoverer, W. E. Hidden. The deep color of this 4.69-carat gem from North Carolina is caused by traces of chromium. Gift of Tiffany & Co. Foundation in 2011.

Topaz

In ancient times, all yellow stones were known as topaz, and it was not until the mid-eighteenth century that the name was linked to the mineral we know as topaz today. The highly prized imperial topaz is intense golden to reddish-orange and is found primarily in Ouro Preto, Brazil. More commonly, topaz is colorless to pale yellow or blue. Since the late 1950s, routine use of radiation and heat treatments that turn pale-colored topaz to deep-blue have provided the buying public with a bountiful supply of affordable blue topaz gems.

OPPOSITE
The selection of topaz gems in various shapes and colors are from Russia, Japan, Madagascar, and Brazil and range in size from 18 to 816 carats.

OPPOSITE
The finest imperial topaz is found near Ouro Preto, Brazil; this 93.6-carat gem from the Washington Roebling collection was cut from a topaz crystal similar to this 875.4-carat one shown here.

ABOVE
This football-shape blue topaz gem from Minas Gerais, Brazil, weighs 7,033 carats. Like most blue topaz, its color is the result of irradiation and heat treatment; originally, this topaz was colorless or pale yellow-brown.

Opal

The name originates from the Sanskrit *upala*, meaning "precious stone." Opal is a noncrystalline form of silica typically found in volcanic rocks or arid sedimentary environments. Gem opal consists of tiny silica spheres (~250 equal the thickness of a sheet of paper) tightly packed together; the voids or spaces between the spheres contain air or water. The play of color in opal is due to the orderly arrangement of these spheres acting like a diffraction grating, breaking visible white light into separate colors. Opal was first discovered in 1872 in Australia, which was the only significant source of precious opal until recent major discoveries in Ethiopia.

The 318.4-carat free-form polished Dark Jubilee Opal, 83.6×62.5×12.5 millimeters (3.29×2.46×0.49 inches), is a black opal from Coober Pedy, Australia. Coober Pedy means "boys' water hole" in a regional aboriginal language. Opals were discovered there in 1915, and to date more than 250,000 shafts have been dug in search of opals. To escape flies and the intense heat, many miners live in underground dugouts. The opal was mined from the fourteen-mile claim and is an approximately one-third portion of the original opal cobble; one section was stolen and the others cut into smaller gems. The Dark Jubilee was cut by Dag Johnson and sold to the Zale Corporation, who donated it to the Smithsonian National Gem Collection in 1980.

ABOVE
Spectacular fire opals from Mexico, ranging from 11 to 143 carats, are sometimes called *lloviznandos*, meaning "drizzle," as they suggest sunlight shining through a Mexican rain shower.

RIGHT
The opal and gold necklace, designed by Louis Comfort Tiffany in about 1915–1925, is accented with Russian green demantoid garnets. The black opals are from Lightning Ridge, Australia. Gift of Mrs. F. R. Downs Jr. and Mrs. R. O. Abbott Jr. in memory of Ruth and Townsend Treadway in 1974.

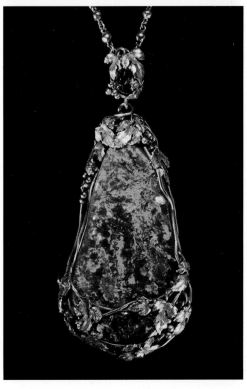

The platinum and gold Opal Peacock Brooch was fabricated by Harry Winston Inc. in 1967 using an opal provided by the client. It features a 30.9-carat black opal from Lightning Ridge, Australia, accented with sapphires, rubies, emeralds, and diamonds. Anonymous gift, 1969.

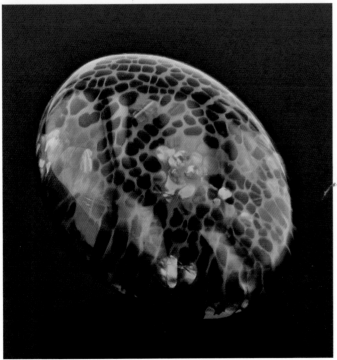

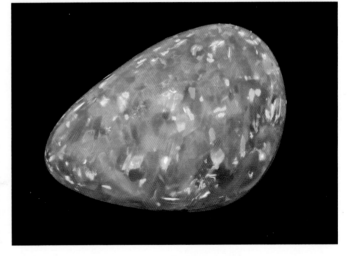

TOP

When cracked or sliced in half, some yowah "nuts," such as this 29.8-carat (size of a Brazil nut) specimen from Queensland, Australia, reveal a valuable gem opal surprise in the center. The donor who contributed the Yowah Opal Nut to the Smithsonian called the specimen his "OMG opal," because every time he would open it up, somebody would say, "Oh, my goodness." Gift of the Richard Ashley Foundation in 2010.

LEFT & ABOVE

During the past decade, Ethiopia has become a new source of spectacular opals, such as these 40.6-carat and 75.3-carat (brown) gems. They are prized for their strong play of color that extends deep into the stones. The larger stone was a gift from Smithsonian Gem and Mineral Collectors in 2016.

The Hope Diamond

Hope Diamond

When Harry Winston gave the Hope Diamond to the Smithsonian Institution in 1958, he imagined it as the cornerstone for building a great national gem collection. He reasoned that even though the United States did not have a king or queen, it should have its crown jewels, and what better place to house them than the Smithsonian Institution in the nation's capital. The Hope Diamond became an immediate star attraction and remains a must-see to this day. True to Winston's dream, it has inspired numerous other gifts of spectacular gems and jewelry that have made the Smithsonian National Gem Collection one of the greatest and most extensive in the world.

Each day, thousands of people crowd into the Hope Diamond gallery to gaze at the famous blue stone. For many, the excitement of finally seeing the iconic gem quickly transitions to questions. The blue color is often a surprise—aren't diamonds colorless? Many are expecting to see the world's largest diamond, perhaps about the size of a baseball, and although 45.52 carats and an inch in diameter is larger than those in a typical engagement ring, it just seems that such a famous diamond should be bigger.

So what has elevated the Hope Diamond to such iconic status? Firstly, its deep-blue color. Blue diamonds are among the rarest objects produced by the earth—only a few are mined each year—and the Hope Diamond is one of the largest and most valuable known. In recent years, blue diamonds have sold for the highest prices at auction of any gems, in some cases approaching $4 million per carat. No doubt, the fact that the Hope Diamond has hosted over two hundred million visitors since taking center stage at the Smithsonian has contributed significantly to its fame. Finally, add a history that includes kings and queens, a daring theft, two recuttings, a wealthy socialite, and according to some, a curse, and you have a provenance that is the envy of any gem.

During the period 1664 to 1668, French merchant and explorer Jean-Baptiste Tavernier completed the last of six expeditions to Persia and India. These were grand affairs involving large entourages and lasting for several years, during which Tavernier and his companions would purchase gems and all variety of exotic goods that were much in demand in Europe. In December 1668, Tavernier showed French king Louis XIV a fabulous collection of diamonds he had acquired during this final trip to India, and in February of the following year the king agreed to purchase them all, including a rare and extraordinary 112³⁄₁₆-carat (115 modern metric carats) blue diamond.

Unfortunately, Tavernier's 1676 published account of his journeys, *Les Six Voyages de Jean-Baptiste Tavernier*, provides no details about when and where he acquired the diamonds, presumably during his last voyage. There is no doubt that the diamonds originated in India, as that was the only commercial source of diamonds in the world at that time (diamonds were discovered in Brazil in 1723 and in South Africa in 1867). There were several diamond mines scattered around India, and although the Golconda region is commonly cited as the source of the blue diamond sold to King Louis XIV, in fact, the mine where it was found, and where Tavernier purchased it, is not known.

ABOVE
Jean-Baptiste Tavernier (1605–1689) was a French explorer and merchant who sold the blue diamond that is the parent stone to the Hope Diamond to King Louis XIV in 1669.

OPPOSITE
King Louis XIV (1638–1718) was the ruler of France from 1643 to 1715. An absolute monarch, Louis XIV believed his power as king was bestowed directly from God. He selected the sun to be his personal emblem and became known as the Sun King. Louis XIV enjoyed a luxurious lifestyle and constructed the magnificent palace and gardens at Versailles.

167

Plate from Tavernier's *Les Six Voyages de Jean-Baptiste Tavernier* showing twenty diamonds sold by Tavernier to King Louis XIV in 1669. The upper left corner (A) shows three views of the blue diamond (~115 metric carats) that was later cut into the French Blue and, ultimately, the Hope Diamond. The plate was rendered by Abraham Bosse, an artist noted for his detailed etchings.

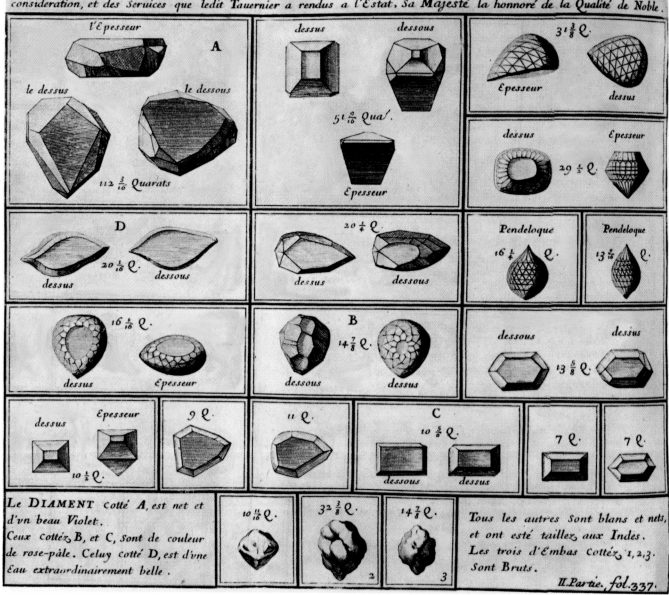

The unmounted Hope Diamond (right) with the lead cast of the French Blue Diamond that was discovered in 2007 in the Muséum National d'Histoire Naturelle in Paris, permitting the first accurate measurements of the size, proportions, and faceting pattern of the historic gem.

King Louis XIV's new blue diamond (known today as the Tavernier Diamond) was cut in the Indian style. Indian diamonds typically were cut to retain size by faceting and polishing faces that followed the natural form of the rough. King Louis XIV tasked one of his court jewelers, Jean Pittan the Younger, with overseeing the recutting of the blue diamond, to enhance its brilliance and achieve a more symmetric shape consistent with contemporary European fashion. The two-year process yielded a 69-metric-carat heart-shape gem, which was referred to as "the great violet diamond of His Majesty" in historical archives. At the time, violet indicated a shade of blue. Today, the diamond is referred to as the French Blue. The 1691 inventory of the French Crown Jewels reveals that the blue diamond was "set into gold and mounted on a stick," a symbol of the king's divinity and power that he could use to impress all who visited Versailles.

Louis XIV's great-grandson King Louis XV commissioned Parisian jeweler Pierre-André Jacquemin in about 1749 to create an emblem of chivalry of the Order of the Golden Fleece. The order was a prestigious fraternity that included many of Europe's crowned heads. The elaborate emblem featured at its center the French Blue Diamond, along with the equally historic 107-carat Côte de Bretagne spinel (then thought to be a ruby) and several other large diamonds. The jeweled emblem likely was seldom worn, serving more as a symbol of the king's wealth and status. In fact, it was typical that royal jewels functioned less as adornments and more as the king's treasury. A strategic display of valuable gems served to intimidate neighboring rivals, and in time of crisis, they were used as collateral to secure quick cash; e.g., to supply an army.

Louis XVI became king in 1775. He and Queen Marie Antoinette were unable to bring about the social and political changes necessary to preserve the monarchy, and when revolution broke out in 1791, they attempted to flee the country but were captured. After the monarchs were arrested, the French Crown Jewels were moved for safekeeping to the Garde-Meuble, a royal storehouse in Paris. In the early morning of September 11, 1792, a group of thieves climbed through the unbarred first-floor window of the Garde-Meuble into the room where the royal jewels were

ABOVE
Color rendering by jeweler Pierre-André Jacquemin of King Louis XV's emblem of the Order of the Golden Fleece (circa 1749). The triangular-shape French Blue Diamond and the 107-carat Côte de Bretagne spinel (carved into the shape of a dragon) are set with several large diamonds.

OPPOSITE
A recent study by Muséum National d'Historie Naturelle and Smithsonian researchers revealed that the central star-facets on the bottom of the French Blue Diamond were cut at an angle such that when looking at the stone set in gold, the effect would be a gold star, or sun, in the center of the blue diamond, as shown in this computer photorealist simulation by François Farges. They concluded that the stone was intentionally cut and set this way by the court jewelers to show the colors of the French monarchy, blue and gold, and the personal emblem of King Louis XIV, the Sun King.

stored and escaped with many of the major gems and jewelry pieces, including the emblem with the French Blue. As the seal on the door of the room had not been broken, and no guards were posted inside the room, the theft went unnoticed. An ever-growing number of thieves returned on subsequent nights and completed the looting of the jewels. Amazingly, the robbery was not discovered until the night of September 17, when a passing patrol noticed the now conspicuous activity around the first-floor window and alerted the guards, who opened the sealed doors to find the room had been stripped bare. Most gems were later recovered in or around Paris, but the French Blue Diamond and the emblem of the Order of the Golden Fleece were never seen again.

Historian Germaine Bapst, in his 1889 *Histoire des Joyaux de la Couronne de France*, concluded that the culprits of the robbery were bands of vagrants who were encouraged by the lack of order in the national guard and had no connection to any of the political factions of the French Revolution. He suggested that one of the thieves, Cadet Guillot Lordonner, took the emblem with the French Blue Diamond on the first night of the robbery and fled with it to London. In 1797, the Côte de Bretagne spinel from the emblem resurfaced in London, and in 1824 was reacquired for the French Crown Jewels (it is exhibited today at the Louvre in Paris).

So, what became of the French Blue Diamond? With the benefit of recent research, we now know that the Hope Diamond is the recut French Blue. The blue color of these diamonds provides a fortuitous advantage for tracking their stories. It is difficult to lose a large blue diamond in history; as they are exceedingly rare, only a few over 20 carats have been found, so their movements are noticed and documented. Colorless diamonds, on the other hand, are relatively plentiful in many sizes, and if stolen or recut simply blend into the crowd of similar looking gems.

Despite the many sensational aspects of the theft of the French Crown Jewels, as there were no documented portraits or drawings of the French Blue Diamond, few people would have known of its existence or details of its appearance. The noted English diamond authority John Mawe in all three editions (1813, 1815, and 1824) of *A Treatise on Diamonds and Precious Stones* describes the great blue diamond *still* in the French Crown Jewels. The first published speculation that the Hope Diamond was the recut French Blue did not appear until 1858 by Charles Barbot—sixty-six years after the theft of the French Crown Jewels.

Unfortunately, history has not yet revealed when and by whom the French Blue was recut into the gem we know today as the Hope Diamond. The first reference to the current diamond is in

Hope's blue diamond is shown in this plate from Edwin William Streeter's *The Great Diamonds of the World: Their History and Romance* set as it was in Hope's collection—in a medallion surrounded by twenty brilliant diamonds with a pearl drop.

an 1812 memorandum written by London jeweler John Francillon. The document and an earlier draft were bound into the 1762 book *Traité des Pierres Précieuses* by Jean Henri Prosper Pouget, which was purchased by noted American gemologist George F. Kunz in London. It was discovered by a Smithsonian researcher in Kunz's papers in the 1970s at the United States Geological Survey Library. The memorandum provides a drawing and description of a blue diamond that matches the weight (177 grains = ~45 metric carats) and shape of the Hope Diamond and indicates that Francillon was shown the gem by diamond merchant Daniel Eliason.

Smithsonian's Mary Winters and John White noted that the date on the Francillon document was twenty years and two days after the theft of the French Crown Jewels, which, interestingly, matched the twenty-year statute of limitations for crimes committed during the French Revolution established by French law in 1804. They suggested that the writing of the memorandum corresponded to the time when the diamond could be openly shown without threat of claim by France. Other scholars have disagreed with this speculation, arguing that the law did not apply to the crown jewels and that the timing was the result of Eliason's financial pressures. Even so, the correspondence of the time periods is striking, and it is possible that the London diamond merchants might have misinterpreted the complexities of the French statutes.

There are several other references to the large blue diamond that mysteriously appeared on the London gem scene. Mawe notes in the 1813 and 1815 editions of *A Treatise on Diamonds and Precious Stones*, "There is at this time a superlatively fine blue diamond, of above 44 carats, in possession of an individual in London, which may be considered as matchless, and of course of arbitrary value." And in his 1823 edition proclaims: "A superlatively fine blue

diamond weighing 44 carats and valued at £30,000, formerly the property of Mr. Eliason, an eminent diamond merchant, is now said to be in the possession of our most gracious sovereign," suggesting that King George IV was in possession of the blue diamond. To this day, however, there has been no evidence found in the British Royal Archives or biographies that link the diamond to King George IV. We do know that an 1839 published catalog by Bernard Hertz documents the diamond as in the collection of London banker Henry Philip Hope.

After Hope's death in 1840, there were extended court proceedings to settle competing claims by his three nephews against his estate, and an 1848 decision awarded the diamond to Henry Thomas Hope. The diamond was bequeathed in 1862 to his wife, Anne Adele Hope, and she named as heir her grandson Francis, who claimed the diamond in 1887. Lord Francis Hope lived extravagantly and accumulated tremendous debt, even more so after he married actress May Yohé in 1894. After years of litigation, in 1901, Lord Francis was granted permission by his family and the courts to sell the Hope Diamond.

The Hope Diamond was purchased in 1901 by diamond merchant Adolf Weil, who the same year sold it to Joseph Frankel's Sons & Company of New York. They quickly realized the challenge of locating a buyer that wanted, and could afford, the magnificent but expensive blue diamond, and by 1907 a depressed diamond market pushed the company to the brink of bankruptcy. They finally sold the stone, at a loss, in 1908, to Turkish diamond collector and dealer Selim Habib, who, because of his own financial troubles, was forced to sell his diamonds at auction in Paris the following year. Before the auction, however, the Hope Diamond was sold separately to French jeweler C. N. Rosenau.

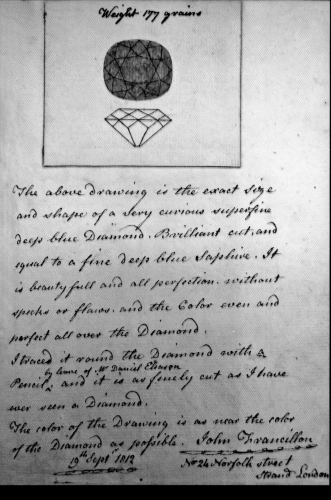

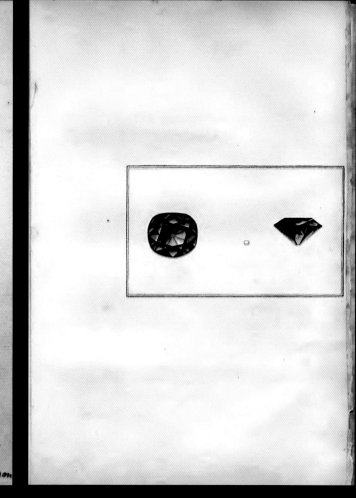

Francillon Memo & Sowerby Etchings

This memorandum written by London jeweler John Francillon in 1812 provides the first historical documentation of the gem we know today as the Hope Diamond. The memo reads:

The above drawing is the exact size and shape of a very curious superfine deep blue Diamond, Brilliant cut, and equal to a fine deep blue Sapphire. It is beauty full and all perfection, without specks or flaws, and the Color even and perfect all over the Diamond.

I traced it round the Diamond with a pencil [by leave of Mr. Daniel Eliason], and it is as finely cut as I have ever seen a diamond.

Also with the Francillon document in Kunz's book is an 1813 pamphlet that describes in French a blue diamond in the possession of Daniel Eliason, with hand-colored etchings of two views of the gem by English naturalist, and son of a lapidary, James Sowerby.

How It Became the Hope Diamond

Gemologist and historian Jack Ogden in the 2019 article "Out of the Blue: The Hope Diamond in London" in *The Journal of Gemmology* offers a possible solution to one of the long-standing mysteries about the Hope Diamond's past. How and when did Henry Philip Hope acquire Eliason's blue diamond? Most chroniclers simply report that the 1839 catalog of Hope's collection included the diamond, with no indication of how it got there. In a marvelous bit of research, Ogden found in the 1845 memoirs of the manufacturer and inventor Sir Edward Thomason his descriptions of the twenty-five most famous diamonds, including one about Blue Hope that recounts the sale of the blue diamond to Hope before mid-July 1821.

Mr. Hope called upon Mr. Elliason [sic] about it, as he had frequently done in admiration of this beautiful gem; but Mr. Elliason always demanded too much. Hearing, however, that it was likely to be hired out for the occasion of the coronation, which circumstance of making it thus public would, in his feelings, much reduce its value, he observed to Mr. Elliason that he called upon him once more respecting the sky blue diamond; and after having stated that he found the King would not purchase it even for the approaching coronation, another opportunity might not occur for years, and he would make him a last offer, conducted, as report says, as follows: Mr. Hope called for pen and ink, and filled up a cheque for 13,000 guineas, placed his watch upon the table, and said he would give Mr. Elliason, five minutes only, to determine to make up his mind, whether to take up the cheque or the diamond. When the time arrived within a few seconds of the five minutes, Mr. Elliason pocketed the cheque, with much grumbling, declaring it more than "dog cheap." Mr. Hope placed the diamond in his splendid collection of minerals among the order of combustibles.

Although a somewhat dramatic telling, the details match those from other sources, and overall, Ogden provides a compelling case that Henry Philip Hope acquired the blue diamond in 1821.

Pierre Cartier purchased the diamond from Rosenau in 1910. He was counting on Cartier's multiple branches and worldwide clientele to locate a suitable buyer, and in fact, he already had one such person in mind. In 1908, he had sold an impressive pearl and emerald necklace and the 95-carat Star of the East Diamond to a young American heiress, Evalyn Walsh McLean, and her husband while they were on their honeymoon in Paris. Cartier showed them the Hope Diamond when they were back in France in 1910, but the couple demurred. Undeterred, Cartier shipped the diamond to New York and had it mounted into a new setting surrounded by sixteen white diamonds. Early in 1911, he brought the necklace to Mrs. McLean's home in Washington, DC, and asked if he could leave it with her over Sunday, as all the banks were closed. In an interview, Evalyn related, "I took it up to bed, and started looking at it on my dresser, and then is when I said I wanted it, after I studied it about half an hour." Cartier's strategy was successful and Evalyn agreed to buy the Hope Diamond for $180,000 (an initial $40,000 plus trade-in of an emerald necklace, and the rest in installments). As it turned out for Cartier, however, the more difficult challenge was receiving payment. It was not until early 1912, after the threat of legal action, that the deal was finally

Smithsonian researchers in collaboration with French colleagues and expert gem cutters used new technology and color information from the Hope Diamond to create a set of accurate cubic zirconia replicas of the Hope Diamond (below, right), as well as the Tavernier (below, left) and French Blue (below, center) Diamonds as they were seen by King Louis XIV. The study concluded that the Hope Diamond is the only surviving piece of the diamond originally sold to King Louis XIV—the rest having been ground away during the various recuttings.

Evalyn Walsh McLean is wearing the Hope Diamond, and suspended beneath it are the 31-carat McLean Diamond and the 95-carat Star of the East Diamond. For a brief period after she purchased the diamond, it was set as the centerpiece of a bandeau-style head ornament, as seen in the photograph on facing page.

Evalyn Walsh McLean

Evalyn Walsh McLean (1886–1947) lived as a young girl in several rough-and-tumble western mining towns. In 1896, her father made a fabulously rich gold strike at the Camp Bird mine in Colorado, and the newly wealthy family moved to Washington, DC, the following year. At age twenty-two, Evalyn married Ned McLean, heir to a newspaper and railway fortune. In 1912, they purchased the Hope Diamond from Pierre Cartier, and it became her signature piece of jewelry. There are numerous photographs of her wearing the Hope Diamond, which is how many people in the United States first became familiar with the gem. The miner's daughter became one of Washington society's best known, and perhaps most eccentric, celebrities. She was famous for hosting extravagant parties where guests were invited to wear the diamond, and on occasion, visitors were greeted by her Great Dane, Mike, with the Hope Diamond dangling from his collar.

As the diamond was with her constantly, security was sometimes improvised, and lax by today's standards. At home she commonly "secured" the diamond under a cushion or pillow. Several years ago, Joseph Bieber related to me a story about his wife's father, Maurice Milstone, who operated a news cart in the early 1920s in front of the *Washington Post* newspaper building, where Evalyn Walsh McLean's husband had his office. On occasion, Evalyn handed him a paper bag, asking him to please hold it for her until she returned in a few days from Miami. At night, Milstone stored the bag at his home, and each day placed it under some papers in his cart. After a few days, Evalyn reappeared and claimed her bag. Bieber's wife recalled that among the jewels in the bag was a velvet pouch containing a splendid diamond necklace with a big blue gem—the Hope Diamond.

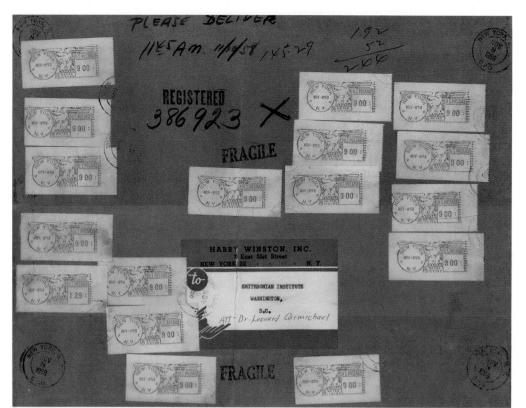

completed. The coming-out party for Mrs. McLean and the Hope Diamond was at an extravagant soiree she hosted for the United States ambassador to Russia on February 2, 1912; the diamond had been remounted as a centerpiece in a bandeau-style headpiece.

The Hope Diamond became Evalyn Walsh McLean's signature gem in Washington's high society. She wore it frequently, layered with her other important gems and jewelry, to lavish parties she famously hosted, as well as when visiting injured soldiers during World War II, where wives and girlfriends were encouraged to try it on. Evalyn Walsh McLean died from pneumonia on April 26, 1947, and two years after her death, her jewelry was sold to pay off debts against her estate. The Hope Diamond, the Star of the East Diamond, and seventy-two other jewelry pieces were purchased by New York jeweler Harry Winston.

Winston included the Hope Diamond in his Court of Jewels, an exhibition of spectacular gems supplemented by a jewelry fashion show that traveled throughout America from 1948 to 1957. In 1958, Winston donated the Hope Diamond to the Smithsonian Institution. On November 10, the Hope arrived at the Smithsonian in a plain brown package shipped by registered mail and insured for $1 million.

As part of the celebration of the fiftieth anniversary of the gift of the Hope Diamond to the Smithsonian, the diamond was temporarily exhibited in this contemporary setting, called Embracing Hope. The necklace consists of three-dimensional ribbons set with 340 baguette-cut diamonds wrapping the Hope Diamond in an exquisite embrace. The necklace was designed by Maurice Galli of Harry Winston and displayed for over a year before being returned to its original Cartier mounting. The Embracing Hope necklace is seen here being worn by Hilary Rhoda.

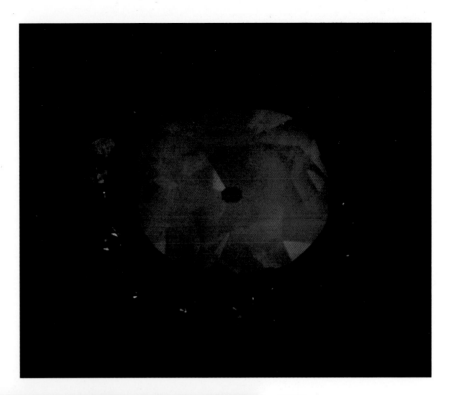

After exposure to ultraviolet light in a dark room, the Hope Diamond emits a strong orange-red light that persists for almost a minute. The cause of the phosphorescence is the same boron impurity that gives the diamond its blue color.

The blue color in the Hope Diamond and others like it is caused by trace amounts of boron. The Hope Diamond is shown here being analyzed using time-of-flight secondary ion mass spectroscopy. The study by Smithsonian researchers determined that on average the Hope Diamond contains about 0.6 parts per million boron.

Is the Hope Diamond Cursed?

Of course not! But where did such a story originate? The "curse" is a late addition to the diamond's history, not making an appearance until the early twentieth century. The first printed mention of the Hope being an unlucky diamond was an account in the *Washington Post* on January 19, 1908 (credited to the *New York Herald*). Titled "Remarkable Jewel a Hoodoo—Hope Diamond Has Brought Trouble to All Who Have Owned It," the article began:

Deep behind the double locked doors hides the Hope diamond. Snug and secure behind time lock and bolt, it rests in its cotton wool nest under many wrappings, in the great vault of the great house of Frankel. Yet not all the locks and bolts and doors ever made by man can ward off its baleful power or screen from its venom those against whom its malign force may be directed.

The article presents a selective and fanciful history of the Hope Diamond that included largely fabricated or exaggerated stories of the misfortunes of former owners of the gem. The most likely explanation for this story is that it was inspired by Frankel's Sons & Company as a last-ditch attempt, along the lines of "any publicity is good publicity," to create interest in the expensive blue diamond that they were desperate to sell. Later that year, their tenuous financial situation forced them to unload the diamond at a loss to Selim Habib, and in its report of the sale, the *Daily Telegraph* continues the curse theme: "The strange story connected with it, that it invariably brings its owners bad luck, is well known." *New York Times* stories chronicling the return to New York in 1910 of the diamond, now owned by Cartier, expanded the rapidly growing list of supposed, and fictitious, misfortunes attributed to the gem. Pierre Cartier likely emphasized certain aspects of the curse stories as a sales tactic when he was trying to sell the Hope Diamond to Ned and Evalyn Walsh McLean. May Yohé, who was once married to Lord Francis Hope, attempted to rejuvenate her acting career in 1912 by performing in London wearing a copy of the Hope Diamond necklace, and in 1921, she coauthored a book and film about the "cursed" Hope Diamond, blaming it for her misfortunes.

Hope Diamond owner Evalyn Walsh McLean often spoke of the curse, at one point seemingly embracing it by asserting that what was bad luck for other people was good luck for her. But events in her life confirmed the diamond's dark powers for many believers. Despite her wealth and publicly glamorous lifestyle, she experienced many tragedies. Her son Vincent was killed at an early age in an automobile accident, her daughter died of a drug overdose, and her husband ended up in a mental asylum; she died, in many ways an unhappy woman, at the relatively young age of sixty-one. How else to explain such terrible things happening to such a privileged person—other than a cursed diamond?

Sue Patch, in *Blue Mystery*, methodically dismisses many of the stories that are commonly invoked to support the existence of a curse on the Hope Diamond. Jean-Baptiste Tavernier was not ripped apart by dogs; in fact, he was made a noble by King Louis XIV and died of an illness in Russia at the ripe old age of eighty-three. There is no evidence that Tavernier's blue diamond ever was part of an Indian idol, and that story echoes those about other gems in various novels and movies. Ill-fated Queen Marie Antoinette's misfortunes are commonly blamed on the cursed diamond, but during her reign, the French Blue Diamond was mounted in the emblem of the Order of the Golden Fleece and only would have been worn by the king (of course, one might argue that his life also did not end well). As with most any object with a long and public history, one can selectively choose, and exaggerate, enough misfortunes to build a good curse story. Considering its arrival inspired unprecedented public interest in the National Museum of Natural History and launched a great National Gem Collection, for the Smithsonian Institution the Hope Diamond has been a wonderful source of good luck!

The 45.52-carat Hope Diamond
removed from its setting. The inset
shows the Hope Diamond's actual size.

Glossary

asterism

Intersecting bright bands on certain polished gems resulting from light reflecting from sets of tiny, parallel needlelike crystals (commonly rutile), which naturally formed inside of a gem crystal. In ruby and sapphire, the threefold symmetry of the corundum atomic structure constrains the rutile crystals to align with equal probability in three directions at 120 degrees to each other, and the resulting three bands of reflected light intersect to form a six-rayed star.

baguette

A small step-cut gem with a long rectangular shape, commonly used as an accessory stone in jewelry.

brilliant cut

The cut most commonly used for diamonds that best balances brilliance and fire. The standard brilliant cut consists of fifty-eight facets. Variations include the marquise, pear (pendeloque), and cushion cut.

briolette

A teardrop- or barrel-shape gem covered with triangular or rectangular facets.

The ivory camels and yellow-gold treasure chests are set with more than one thousand gems—round brilliant-cut diamonds and cabochons of ruby, sapphire, and emerald—and likely were fabricated in Hong Kong in the 1950s for the high-end tourist market. The treasure chests mounted on either side of the carved ivory camels open to reveal gem "treasures." Gift of Mr. George H. Capps in 1980.

cabochon

A gem cut with a smooth, rounded top surface, typically used for opaque gems or to show off effects, such as stars or cat's eyes.

cat's eye

A single bright line on the surface of a gem caused by reflection of light from tiny, parallel needlelike crystals that formed in the original gem crystal. Cat's eyes are best seen on polished rounded surfaces (cabochon) and are most commonly associated with chrysoberyl gems.

carat

Typically, gemstones are measured by their weight in carats. The metric carat is defined as 0.2 grams (.007 ounces), and each carat is divided into one hundred points. The word "carat" comes from carob, the name of a tree that grows in the Mediterranean region. For centuries, carob seeds were used as a standard for weighing precious stones. The metric carat was adopted in the United States in 1913 and throughout the world by 1930. Prior to that time, many different definitions of carat were used, making for considerable confusion in the gem trade.

clarity

Refers to the presence, or lack, of inclusions or other internal imperfections that affect a gem's overall appearance.

crystal

The natural form of most minerals, meaning it is constructed of atoms that are locked into a precise symmetric pattern that is repeated in an orderly way in three dimensions billions of times. Crystals with favorable conditions, and room to grow, might form spectacular geometric shapes bound by smooth flat faces. Large, flawless crystals might be cut into gems.

dispersion

The ability of a gem to separate light into its spectral colors. Each color of light travels at a slightly different speed and refracts, or bends, slightly differently as it enters a gem. As the light travels through the gem, the colors become more and more separated, or dispersed. This spread of light into its component colors that change and flash as the gem is moved and tilted is called dispersion, or more commonly referred to as "fire."

fluorescence

Emission of colored light triggered by ultraviolet light. The ultraviolet light temporarily boosts electrons into higher energy levels, and when the electrons fall back into their original levels, they release their excess energy as visible light. If the electrons remain at the higher level for some extended period, the delayed light emission is called phosphorescence.

gem

A mineral crystal that has been cut and polished into an object of great beauty.

marquise

A brilliant-cut gem with an elongated boat shape.

mineral

Solid, typically inorganic, chemical compound that forms naturally in the earth. Minerals are the building blocks of the solid earth; all rocks are made of minerals.

old mine cut

An antique diamond cut that was common during the nineteenth and early twentieth centuries. It is characterized by an uneven shape, large culet (flat bottom facet), and small table (top facet).

pavé

A type of setting in which small faceted stones are inlaid into a piece of jewelry with their tops flush with the surface.

play of color

The flashes of color exhibited by precious opals, caused by diffraction (splitting light into its various colors) of light by the orderly three-dimensional arrangement of silica spheres that make their internal structures.

pleochroism

Light traveling in different directions through some crystals and gems interacts differently with the atomic structures, resulting in distinct colors when viewed in various orientations.

rose cut

An antique diamond cut named for the flower the stones resemble. They have a flat bottom and faceted domed top.

Photography Credits

PAGES 2 AND 131 (TOP RIGHT): Yogo Flower Brooch, courtesy John T. Haynes Inc.

PAGES 4 AND 95: Dom Pedro Aquamarine, Donald E. Hurlbert, Smithsonian Institution

PAGE 7: Group of diamonds, Chip Clark, Smithsonian Institution, enhanced by SquareMoose Inc.

PAGE 8: Gem Gallery, Chip Clark, Smithsonian Institution

PAGE 9: (Top) labradorite sunstone, Ken Larsen, Smithsonian Institution; (bottom) jadeite cobble, Chip Clark, Smithsonian Institution

PAGES 10 AND 20 (LEFT): Model wearing diadem with emeralds, © the Estate of Erwin Blumenfeld

PAGE 12: Portrait of Marjorie Merriweather Post, Hillwood Estate, Museum & Gardens Archives

PAGE 13: Ms. Post presenting Napoleon Diamond Necklace, Hillwood Estate, Museum & Gardens Archives

PAGE 14: Portrait of Marie Antoinette, © RMN-Grand Palais/Art Resource, NY

PAGE 15: Marie Antoinette Diamond Earrings, Chip Clark, Smithsonian Institution, enhanced by SquareMoose Inc.

PAGE 16: Portrait of Princess Tatiana Alexandrovna Yusupov, courtesy State Museum of the History of St. Petersburg

PAGE 17: Diamonds out of settings, Chip Clark, Smithsonian Institution

PAGE 18: Marjorie Merriweather Post wearing diadem, Hillwood Estate, Museum & Gardens Archives

PAGE 19: Diadem with turquoise, Chip Clark, Smithsonian Institution, enhanced by SquareMoose Inc.

PAGE 23: Napoleon Diamond Necklace, Chip Clark, Smithsonian Institution, enhanced by SquareMoose Inc.

PAGE 24: (Top) Napoleon Diamond Necklace detail, Chip Clark, Smithsonian Institution; (bottom) original leather case, Ken Larsen, Smithsonian Institution

PAGE 25: Archduchess Maria Theresa, courtesy Osterreichische Nationalbibliothek Bildarchiv, Vienna

PAGE 26: Blue Heart Diamond and Hope Diamond, Chip Clark, Smithsonian Institution

PAGE 27: Blue Heart Diamond ring, Chip Clark, Smithsonian Institution, enhanced by SquareMoose Inc.

PAGE 28: (Left) Blue Heart Diamond in lily-of-the-valley corsage ornament, Cartier Archives, Paris, © Cartier; (right) photograph of Maria Unzue de Alvear, © RMN-Grand Palais/Art Resource, NY

PAGE 29: (Left) photograph of Nina Dyer wearing Blue Heart Diamond pendant, courtesy Vincent Meylan; (right) VCA drawing of Blue Heart Diamond pendant, Van Cleef & Arpels Archives, © Van Cleef & Arpels SA

PAGE 30: Portrait of Maharaja of Rajpipla, collection of Prince Indra Vikram Singh of Rajpipla

PAGE 31: Post Emerald Necklace, Chip Clark, Smithsonian Institution, enhanced by SquareMoose Inc.

PAGE 32: Ms. Post wearing emerald necklace, Hillwood Estate, Museum & Gardens Archives

PAGE 33: (Bottom) Post Emerald Necklace detail, Chip Clark, Smithsonian Institution; (top) photograph of Ms. Post dressed as Juliet, Hillwood Estate, Museum & Gardens Archives

PAGE 35: Maximilian Emerald ring, Chip Clark, Smithsonian Institution, enhanced by SquareMoose Inc.

PAGE 36: Post Diamond Tiara, Laurie Minor-Penland, Smithsonian Institution

PAGE 37: Jadeite Dragon Vase, Chip Clark, Smithsonian Institution

PAGE 39: Logan Sapphire, Chip Clark, Smithsonian Institution, enhanced by SquareMoose Inc.

PAGE 41: (Left) Logan Sapphire fluorescing, Jeffrey Post, Smithsonian Institution; (right) Polly Logan at the Smithsonian, Smithsonian Institution Archives

PAGE 43: Bismarck Sapphire necklace, Chip Clark, Smithsonian Institution, enhanced by SquareMoose Inc.

PAGE 44: Photograph of Mona von Bismarck, Getty Images

PAGE 45: Photograph of Bismarck Sapphire necklace, Cartier Archives, New York, © Cartier

PAGE 47: Rosser Reeves Star Ruby, Chip Clark, Smithsonian Institution, enhanced by SquareMoose Inc.

PAGE 143: Group of quartz gems, Chip Clark, Smithsonian Institution, enhanced by SquareMoose Inc.

PAGE 144: (Top) Angelina Jolie wearing the Jolie Citrine Necklace, courtesy Robert Procop; (bottom) Tiffany amethyst necklace, Chip Clark, Smithsonian Institution

PAGE 145: Faceted quartz egg, Chip Clark, Smithsonian Institution

PAGE 146: (Top) amethyst gem, Priscilla Dyer; (bottom left and right) iris agate and agate slice, Chip Clark, Smithsonian Institution

PAGE 147: (Top left) rose quartz on quartz, Chip Clark, Smithsonian Institution; (bottom left) Nevada amethyst, Greg Polley, Smithsonian Institution; (right) Dyber ametrine, Sena Dyber

PAGE 148: Garnet hairpin, Chip Clark, Smithsonian Institution

PAGE 149: (Top) group of garnet gems, Greg Polley, Smithsonian Institution; (bottom) tsavorite garnet, Chip Clark, Smithsonian Institution

PAGE 150: Zircon gem, Greg Polley, Smithsonian Institution

PAGE 151: (Top) George Zircon Brooch, Greg Polley, Smithsonian Institution; (bottom) group of zircon gems, Dane Penland, Smithsonian

PAGE 152: Spinel crystal on matrix, Federico Bärlocher

PAGE 153: (Top) spinel bracelet, Chip Clark, Smithsonian Institution; (middle left) spinel ring and (bottom right) pink spinel, Greg Polley, Smithsonian Institution; (bottom left) blue spinel, Chip Clark, Smithsonian Institution

PAGE 154: Lotus kunzite gem, photograph by Arjuna Irsutti, courtesy Victor Tuzlukov

PAGE 155: (Top) kunzite heart gem, Laurie Minor-Penland, Smithsonian Institution; (bottom left) spodumene gem and (bottom right) Oceanview kunzite gem, Greg Polley, Smithsonian Institution; (middle right) hiddenite gem, Ken Larsen, Smithsonian Institution

PAGE 157: Group of topaz gems, Chip Clark, Smithsonian Institution, enhanced by SquareMoose Inc.

PAGE 158: Imperial topaz gem and crystal, Chip Clark, Smithsonian Institution, enhanced by SquareMoose Inc.

PAGE 159: Blue topaz gem, Laurie Minor-Penland, Smithsonian Institution

PAGE 160: Dark Jubilee Opal, Chip Clark, Smithsonian Institution

PAGE 161: (Top) group of Mexican opals and (bottom) Tiffany opal necklace, Chip Clark, Smithsonian Institution

PAGE 162: Opal Peacock Brooch, Chip Clark, Smithsonian Institution, enhanced by SquareMoose Inc.

PAGE 163: (Top) Yowah Opal Nut, Smithsonian Institution; (bottom left and right) Ethiopian opals, Greg Polley, Smithsonian Institution

PAGE 164: Hope Diamond necklace, Dane Penland, Smithsonian Institution, enhanced by SquareMoose Inc.

PAGE 169: Lead cast of French Blue with the Hope Diamond, François Farges

PAGE 171: Photorealistic image of French Blue, François Farges

PAGE 173: Francillon memo, United States Geological Survey

PAGE 174: Henry Philip Hope, © National Portrait Gallery, London

PAGE 175: Replicas of Tavernier, French Blue, and Hope Diamonds, photograph © 2018 John Bigelow Taylor, courtesy John Hatleberg

PAGE 176: Evalyn Walsh McLean, Smithsonian Institution Archives

PAGE 178: (Top) Registered Mail package for the Hope Diamond, Smithsonian Institution Archives

PAGE 179: Hope Diamond presented to the Smithsonian by Mrs. Winston, Smithsonian Institution Archives

PAGE 180: Embracing Hope Necklace worn by Hilary Rhoda, © Robbie Fimmano

PAGE 181: (Top) Hope Diamond phosphorescing, Chip Clark, Smithsonian Institution; (bottom) Hope Diamond in TOF-SIMS, Jeffrey Post, Smithsonian Institution

PAGE 183: Hope Diamond, Chip Clark, Smithsonian Institution

PAGE 184: Ivory camels, Ken Larsen, Smithsonian Institution

Index

This book is dedicated to the many people whose generous gifts have built the Smithsonian National Gem Collection.

Acknowledgments

This book would not have been possible without the support and assistance of many talented colleagues and collaborators. Russell Feather, collection manager of the Smithsonian National Gem Collection, skilled gemologist and relentless gem sleuth, was a critical source of information for many of the gem stories told here. Christine Webb was an enthusiastic partner, and friend, in all aspects of preparing this book, including research, organizing images, and reviewing text. Her experience and knowledge of gems, and love of jewelry, were invaluable. I received essential, and always cheerful, assistance from Kealy Gordon and Adam Mansur with images and overall logistics related to creating this book. The research for this book was greatly assisted by Holly Heighes, who expertly, and always with good humor, organized decades of paper files into a digital archive. It was a pleasure and my privilege to work with all these special people.

Jill Corcoran of Smithsonian Enterprises was an early advocate for this book, and I greatly benefited from her guidance and moral support throughout the process. I very much enjoyed my collaboration with editorial director Shawna Mullen at Abrams. Her skills, good-natured guidance, and love for gems greatly improved the manuscript and made the work on this book interesting and enjoyable. I am grateful to designer Eli Mock for his patience and creativity in combining text and numerous images into a beautiful book.

I am grateful for the encouragement and friendship of Loretta Cooper and the members of the Smithsonian Gem and Mineral Collectors, and for their passion for sharing earth's great treasures with the people of the world. Finally, this book would not have been completed without the love and support of my wife, Ann, and my daughters, Alison and Marissa. Ann reviewed text, and her helpful guidance and encouragement greatly improved the book. They are the true "gems" in my life!

Editor: Shawna Mullen
Designer: Eli Mock
Production Manager: Kathleen Gaffney

Library of Congress Control Number: 2020944159

ISBN: 978-1-4197-4580-5
eISBN: 978-1-68335-940-1

Printed and bound in Malaysia
10 9 8 7 6 5 4 3

Abrams books are available at special discounts when purchased
in quantity for premiums and promotions as well as fundraising
or educational use. Special editions can also be created to speci-
fication. For details, contact specialsales@abramsbooks.com or
the address below.

Abrams® is a registered trademark of Harry N. Abrams, Inc.

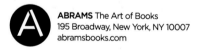

ABRAMS The Art of Books
195 Broadway, New York, NY 10007
abramsbooks.com